Unlocking the Metaverse

Unlocking the Metaverse

A Strategic Guide for the Future of the Built Environment

Paul Doherty

Collierville
Tennessee, USA

Published by John Wiley & Sons, Inc., Hoboken, New Jersey.
Published simultaneously in Canada.

For general information on our other products and services or for technical support, please contact our Customer Care Department within the United States at (800) 762-2974, outside the United States at (317) 572-3993 or fax (317) 572-4002.

Wiley also publishes its books in a variety of electronic formats. Some content that appears in print may not be available in electronic formats. For more information about Wiley products, visit our web site at www.wiley.com.

Library of Congress Cataloging-in-Publication Data applied for

Hardback ISBN: 9781394198764

Cover Design: Wiley
Cover Image: © koiguo/Getty Images

Set in 10.5/13pts ChapparalPro Regular by Straive, Chennai, India

SKY10059348_110623

Contents

About the Author

Paul is chairman and CEO of TDG (the digit group www.thedigitgroupinc.com) and NextGEN Entertainment, Inc. (www.nextgenentertain.com), and is a globally renowned and award-winning architect who is one of the world's most sought after thought leaders, strategists, and integrators of process, technology, and business. As noted in *The Wall Street Journal*, seen on Bloomberg TV, acknowledged by CNBC as one of America's Business Titans, and reported by Forbes as "Changing the World," Paul is a Senior Fellow of the Design Futures Council and a Fellow of the International Facility Management Association (IFMA). Paul currently works as chairman of TDG Global Ventures, a Smart City real estate development company that provides master planning, access to financing, program management, and innovative technology solutions around the world. Concurrently, Paul is the co-founder and producer of the critically acclaimed AEC Hackathon (www.aechackathon.com) that launched at Facebook Headquarters. His past successful ventures include Revit Technologies (Sold to Autodesk 2002), Buzzsaw (Autodesk 2001), and TRIRIGA (IBM 2011).

Acknowledgments

I would like to acknowledge the following people as this book would never have been able to be delivered without their love and support. My wife Jessie Doherty, my son Daniel Doherty, my mother Andrea Doherty, my sister Erin, and her family.

I want to thank the talented team at my publisher Wiley, Kalli Schultea, Indirakumari S., and Isabella Proietti, and I am sure there is a team of people who I have not mentioned who made this book process and production a valuable exercise.

I also thank my partners, colleagues, and friends V. Marbue Dennis III, Arol Wolford, Amit Chopra, Steve Jones, David Uslan, Michael Uslan, Nancy Uslan, Tarek Abbas, Amr Attar, Rabelin Tchoumi, Gordon Cheng, Margie Petherick, Thomas Doherty, Pierre Lo, Bill Wang, Dr. Turki Shoaib, Dr. Karen Stephenson, Wyly Wade, Cody Nowak, Terry Beaubois, Patrick Mays, Alain Waha, Dimitri Vegas & Like Mike, Hidetoshi Dote, Patrick Sharpe, Normandy Madden, Joe Montgomery, Dave Gilmore, James Cramer, Silvia Davidia, Remi Arnaud, Matthew Tribe, Ala Hassan, Hugh Seaton, Damon Hernandez, Chase Olson, Joshua Gumbiner, Don Bowden, Matt Abeles, Cristina Savian, Christina Lu, Dr. Caroline Chung, Dr. Anas Bataw, and the Trustees of Humanity – all of whom deserve so much thanks and love.

Acknowledgements

List of Acronyms

AEC	architecture, engineering, construction
AI	artificial intelligence
API	application programming interface
AR	augmented reality
BIM	building information modeling
BPR	business process re-engineering
CAD	computer-aided design
DLT	distributed ledger technology
ESG	Environment Sustainable Governance
ETF	exchange traded funds
FM	facility management
GPT	generative pre-trained transformer
IBM	international business machines
ICT	information communications technology
IoT	Internet of Things
IP	internet protocol
IRL	in real life
OEM	other equipment manufacturers
TDG	The Digit Group, Inc.
VPN	virtual private network
VR	virtual reality
XR	eXtended reality and/or mixed reality

Introduction: How to Use This Book

This introduction will provide an overview and road map of the book's content to allow readers to have a clear understanding of how to use this book's information for their own valued use.

Unlocking the Metaverse

Welcome to *Unlocking the Metaverse: A Strategic Guide for the Future of the Built Environment*. When the physical built environment intersects with the digital world, not only is it a moment to pay attention to, but it's time to write a book about this historic moment. Welcome to a journey of unlocking the mysteries of the metaverse for the built environment and how these technologies, processes, workflows, and experiences will affect you, your business, and your lifestyle. My goals of writing this book are:

- To provide the background of how the definitions of our age are becoming our new reality.
- To provide the background and path forward of how our 3D graphic representation of the built world and its incorporation into storytelling has given rise to an enormous and influential Gaming Industry and how this affects the built environment.
- To expose opportunities to the incumbent and traditional professionals of the built environment as well as the cautionary tales of what the digital asset world has unleashed.
- To provide guidance and suggestions for how you, your company, and your lifestyle will be affected by the new metaverse medium.

- To suggest practical advice on emerging data-driven innovations like Distributed Ledger Technology (Blockchain), Smart Contracts, and Tokenomics will give rise to the next generation of valuable digital real estate.
- To create vision to the emergence of Web3, Artificial Intelligence and when integrated into Metaverse environments, how it transforms the notion of a Cyber–Physical experience.
- And finally, to bring views from industry experts to look into the Crystal Ball to assist you in navigating this amazing transformative age we are experiencing (Figure 1).

My first experience of the metaverse was 1977, playing with my friend's Atari 2600 gaming console on a game called Pong. I was transfixed with seeing a home video game on a TV. By 1979, we had graduated to games like Asteroids, Space Invaders, and the beginnings of digital football, baseball, and ice hockey. I was not a big fan of the arcades like pinball games and other console games, but Pac-Man and Ms. Pac-Man were always fun and for a short period of time, allowed a form of escapism. Thus, my take that these home video games and arcades from the 1970s and 1980s were seeds for the promise of the oncoming of a new medium called the Metaverse.

Fig. 1: Expo2020 in Dubai Metaverse.
Courtesy: Paul Doherty.

Fig. 2: Oracle Park in the Metaverse, home of the San Francisco Giants baseball team. Courtesy: Paul Doherty.

As computing power increased and storytelling matured, I was enamored in 1993 with the multimedia adventure game called Myst, by Broderbund. I used my Apple Macintosh with a Myst CD-ROM disk to propel me into a fictional world where I would spend hours unlocking mysterious clues in a never-ending game. The 1990s were full of milestones for the maturity of the Metaverse, with the rise of the Internet and its adoption in the later part of that decade and Neal Stephenson coining the term "Metaverse" in his cyberpunk novel *Snow Crash* in 1992 to describe an online, Virtual Reality (VR) world where the inhabitants of humankind could interact and escape the dystopian unpleasantness of the real world (Figure 2).

In fact, in 1996, I was an advisor to companies that were working on the new San Francisco Giants baseball stadium. The designers wanted to have stakeholders experience what the views were from seats in the new stadium before it was built and did not want to purchase expensive design software just for viewing the 3D model. Our team, led by Planet9, worked with a new technology for its time called Virtual Reality Markup Language (VRML) that allowed the 3D model of the new stadium to be viewed and interacted with inside a Web Browser. I documented the process and images in my 1997 book called *Cyberplaces: The*

Fig. 3: Cyberplaces book cover.
Courtesy: Paul Doherty.

Internet Guide for Architects, Engineers and Contractors published by R.S. Means. When I hear about people claiming that today they are creating the Metaverse or are self-proclaimed experts regarding the Metaverse, I shake my head knowing that the world has many charlatans, as I literally wrote the book on the emerging medium of the metaverse in 1997. This validation is brought forward into today's world through the valuable work of so many talented people over the decades that has led us collectively to this moment in time and with you reading this book (Figure 3).

The metaverse at this moment in the time of the writing of this book is not mature enough to say that it has arrived, but rather pieces of this new medium are being put in place as technology and our use of it matures. Many of the elements of the metaverse are being defined, developed, and delivered in a velocity that I am surprised by its speed. I like to think of these elements as gemstones that as the metaverse matures as a strand to tie them together like a beautiful necklace.

Like an elephant, the metaverse is being described by people with them touching only one part of the elephant, we also must understand who else is touching that elephant and hear their stories before the complete elephant is identified. By reading this book, you have the opportunity to properly define our new Age, by thinking, doing, implementing, and sharing.

The Method Behind This Book

With this background in mind, the methodology of the use of this book is multi-dimensional. I use both a first person and observer viewpoint within each chapter to provide you with both a strategic and practical process for how to best implement your own approach to the metaverse. Behind each chapter, I have also provided real-world stories as examples of how the emerging elements of the metaverse are being planned, implemented, and measured. The multi-dimensional nature of this book is based on how different stakeholders in the built environment are directly affected by the metaverse, which resulted in a book that attempts to showcase how our new immersive, human-centric but data-driven work can be used by the average human on our planet earth.

Who is this Book For?

This book is written to be both reference material and as a guide for our broad industry. Designers, architects, and engineers will find value in a good portion of this book, while contractors and the trades will find value in the implementation areas of this book. Facility managers and operators will discover value mechanisms and measures from this book that may change how their work gets done, while the common reader will gain insight that the metaverse as it transforms into an Industrial Metaverse, can be of value to them over today with an eye towards a decentralized industry and society in the future.

The Book Road Map

This introduction chapter is meant to be a reference that provides the proper context for you and acts as a foundation for your metaverse journey.

Chapter 1: Definitions will provide you with the basic terms and descriptions of our new Age of the Metaverse.

Chapter 2: Digital Twins, Virtual Worlds, and the Metaverse gives insight into how these new visual mediums will affect the different areas of the built environment from the perspectives of pre-design, design, construction, operations, and maintenance.

Chapter 3: Metaverse Mechanisms and Solutions explores the new data algorithms, structures, and processes that leverage distributed ledger technology (DLT) in various forms like blockchain, cryptocurrency, smart contracts, and digital real estate valuations. This chapter also explores innovations that provides an initial peak into the rapidly emerging next generation of the World Wide Web in the form of Web 3 and its ability to host powerful functions like artificial intelligence (AI) and generative pre-trained transformer (GPT) technologies that promise to change our industry from its traditional processes into new ways of working that one can only begin to imagine.

Chapter 4: The Crystal Ball puts you in the middle of conversations about our collective future as an industry. Highlighted with interviews of some of the leading voices in the world today concerning technology and the built environment, this chapter provides inspiration and wonder as to what the future holds for all of us (Figure 4).

Fig. 4: Author Paul Doherty.
Courtesy: Paul Doherty.

As you read through this book, I encourage you to interact with me to challenge and contribute our concepts and solutions as the world changes and moves forward. I look forward to your communication and constructive criticism on social media. In the meantime, please enjoy my book as a milestone moment in time, not as an absolute.

Welcome to the world of the metaverse!
Paul Doherty
Memphis, Tennessee, 2023

CHAPTER

1

Definitions

The introduction of new technologies into the built environment and its professions are fraught with the challenges of hype, limited adoption, and, in some cases, indifference. In some cases, there is a trend that new technologies are showing signs of adoption but need guidance, education, and sound strategies in order to become valuable to the built environment industries and professions. A first step in the education and communication process is to define the numerous elements that are affecting our built environment. The following definitions are a snapshot from this moment, a foundation to build upon and are meant to be revisited and challenged as we learn, fail, succeed, and adapt to our ever-changing world of the intersection of the physical and digital worlds.

The following definitions are not meant to be comprehensive, but rather to be used as a referenced foundation for your journey of moving through the rest of this book.

Metaverse

William Gibson's quasi-prophetic vision of cyberspace as a "consensual hallucination" manifests itself in the emerging definition of the metaverse. Our company describes metaverse as an analogy to Outer Space. There is no there, there. It is a digital immersive environment that has the equivalent of galaxies, solar systems, and planets (virtual worlds) that provide a sense of place. The term "metaverse" originated in the 1992 science fiction novel *Snow Crash* as a portmanteau of "meta" and "universe." Metaverse development is often linked to advancing virtual reality technology due to the increasing demands for immersion. Recent interest in metaverse development is influenced by Web3, a concept for a decentralized iteration of the internet. Web3 and metaverse have been used as buzzwords to exaggerate the development progress of various related technologies and projects for public relations purposes. Information privacy, user addiction, and user safety are concerns within the metaverse, stemming from challenges facing the social media and video game industries as a whole. What we are learning as a company about metaverse is that through iterative design and delivery of 3D immersive experiences, is that new measures of usage and value need to be developed and implemented. We use terms such as Level of Experience and Level of Engagement (LOE) to describe how a user of the metaverse is immersed in a 3D digital environment.

At its current core, the metaverse is a spatial computing platform that provides digital experiences as an alternative to or a replica of the real world, along with its key civilizational aspects like social interactions, currency, trade, economy, and property ownership, all experienced on a bedrock of blockchain technology. Metaverse is an interaction, and, in certain cases, an integration of 3D worlds accessed through a browser, a mobile app, or a headset. It allows people to have real-time interactions and experiences across large distances, either by themselves, one-on-one and/or with many people. A vast ecosystem of online applications is emerging to build these human and non-human interactions, communications, and relationships (Figure 1-1).

The metaverse is an environment where the physical and digital worlds can coexist and significantly impact fundamental areas of daily life. It is a universe of limitless, interconnected virtual communities where people can socialize, collaborate, and have fun. It may include additional aspects of online life, such as social media and shopping. As application scenarios mature, the metaverse will develop into an exceptionally large-scale, extremely open, and dynamically optimized system. To create a system that can support various virtual reality application scenarios, creators from different fields will need to work together to fulfill the promise of the metaverse.

Fig. 1-1: Expo2020 Metaverse in Dubai.
Courtesy: Paul Doherty.

Our company is partnering and strategizing with a broad range of creators to deliver our virtual worlds in the metaverse around the theory that Web3 will be the next generation of the Internet as a 3D immersive environment that links virtual worlds in the metaverse together similar to how hyperlinks work in the current Web2 Internet environment of 2D web sites. Immersive worlds with next-generation hyperlinks are an amazing world to explore.

In the context of the built environment, the metaverse can be used to visualize and manage various aspects of a physical facility, such as building layouts, maintenance schedules, and resource allocation. For example, a facility manager could use a virtual reality (VR) headset to explore and interact with a 3D model of a building, viewing detailed information about various systems and equipment, such as HVAC, plumbing, and electrical. They could also use the metaverse to track and schedule maintenance tasks, assign work orders to technicians, and track the progress of ongoing projects.

Another example is through the use of augmented reality (AR) technology. A facility manager could use an AR app on their smartphone or tablet to view information about a specific location or piece of equipment with a digital overlay on their physical surroundings. This could be useful for accessing technical manuals, viewing repair histories, or identifying potential issues. The use of the metaverse in facility management can help improve efficiency, reduce costs, and improve the overall management of physical facilities (Figure 1-2).

Fig. 1-2: Qingdao, China Virtual Reality Theme Park Metaverse.
Courtesy: Paul Doherty.

Digital Twin

A digital twin is a digital representation of a physical object or system. It is typically created using data from sensors, simulations, and other sources, and it is used to analyze, optimize, and predict the behavior and performance of the corresponding physical object or system. Digital twins are often created using software tools such as computer-aided design (CAD) and Building Information Model (BIM) programs, and they can be updated in real time as new data becomes available. They can also be integrated with other systems, such as manufacturing equipment or transportation networks, to enable real-time monitoring and control.

Digital twins can be used to optimize the built environment by providing a virtual model of the facility that can be used to simulate and analyze different scenarios, such as predicting maintenance needs or optimizing energy use. Here are some examples of how digital twins can be used in facility management:

- Predictive maintenance: By continuously monitoring the performance of physical assets and comparing it to the digital twin model, facility managers can predict when maintenance is needed and schedule it in advance, reducing downtime and improving efficiency.
- Energy optimization: Digital twins can be used to simulate different energy consumption scenarios and optimize energy use in a facility. For example, a digital twin of an office building

could be used to determine the most energy-efficient lighting and temperature settings.

- Space optimization: Digital twins can help facility managers optimize the use of space in a facility. For example, a digital twin of a warehouse could be used to simulate different layouts and determine the most efficient use of space.
- Emergency management: Digital twins can be used to simulate different emergency scenarios and develop emergency response plans. For example, a digital twin of a hospital could be used to simulate a fire and determine the best evacuation routes and resources needed to respond.
- Virtual commissioning: Digital twins can be used to test and commission new facilities or equipment before they are built or installed, reducing the need for physical testing, and improving efficiency.

When used in construction, digital twins can refer to digital replicas of physical assets and systems used to simulate, monitor and optimize their performance. Some examples of digital twins in construction:

- Building Information Modeling (BIM): BIM is a digital twin of a building, which provides a 3D model of the building's physical and functional characteristics. It can be used to simulate and optimize the building's performance, energy efficiency, and safety.
- Construction site simulation: Digital twins can be used to simulate construction sites to optimize the construction process and improve safety. For example, a digital twin of a construction site can be used to simulate the movement of materials and workers, identify potential safety hazards, and optimize the layout of the site.
- Smart infrastructure: Digital twins can be used to monitor and optimize the performance of infrastructure assets such as bridges, roads, and tunnels. For example, a digital twin of a bridge can be used to monitor its structural health, predict maintenance needs, and optimize traffic flow.
- Asset management: Digital twins can be used to manage and optimize the lifecycle of construction assets such as cranes, excavators, and trucks. For example, a digital twin of a crane can be used to monitor its performance, predict maintenance needs, and optimize its usage.

Virtual Worlds

Virtual worlds are 3D digital environments or spaces that exist online and can be accessed and interacted with through the internet. They can be used for a variety of purposes, such as socializing, entertainment, education, and business. Virtual worlds can be accessed through a variety of devices, including computers, smartphones, and virtual reality headsets. They can take the form of social media platforms, online multiplayer games, virtual reality environments, or virtual-reality-based social spaces. In a virtual world, users can create and customize their own digital avatars, or online identities, which they use to interact with other users and the environment. Users can also participate in activities and events, communicate with other users through text or voice chat, and explore different areas and features within the virtual world.

Virtual worlds can be designed to mimic real-world environments or can be entirely fictional and imaginative. They can be used for a wide range of purposes, including socializing with friends, playing games, attending virtual events and concerts, participating in educational or training programs, and even conducting business meetings and transactions (Figure 1-3).

Fig. 1-3: A virtual world in the Metaverse.
Courtesy: Paul Doherty.

Virtual Worlds are the provenance of the Industrial Metaverse that can build upon a vertical business sector and focus its data from this vertical viewpoint. This approach promises to provide value to a specific industry faster than waiting for the digital "big bang" to occur on a global, multi-sector scale.

Blockchain

A blockchain is a decentralized, distributed ledger that is used to record transactions across many computers so that the record cannot be altered retroactively without the alteration of all subsequent blocks and the consensus of the network. Each block in the chain contains a record of multiple transactions, and once a block is added to the chain it cannot be altered. The decentralized nature of a blockchain makes it resistant to fraud and tampering, as there is no single point of control.

Blockchains are often associated with cryptocurrencies, such as Bitcoin and Ether, but they can be used to record and verify any type of data or information. They have the potential to revolutionize our industry by providing a secure, transparent, and immutable record of transactions and other data. Some examples of how blockchain technology is being used in the real estate industry:

- Property ownership and transfer: Blockchain can be used to record and verify ownership of real estate assets, making it easier to track the ownership of a property and facilitate the transfer of ownership when the property is sold.
- Mortgage and loan origination: Blockchain can be used to securely store and verify mortgage and loan documents, streamlining the loan origination process and reducing the risk of fraud.
- Lease management: Blockchain can be used to track and verify rental agreements, making it easier for landlords and tenants to manage their leases and ensure that all parties are following the terms of the agreement.
- Facility management: Blockchain can be used to track and verify facility and property maintenance and repair records, making it easier for property managers to manage the upkeep of their properties.
- Land registry: Blockchain can be used to maintain a secure and transparent record of land ownership, making it easier for governments and other organizations to track and verify land ownership and reduce the risk of fraud.

Fungible Token (FT)

A fungible token is a type of digital asset that is interchangeable and identical in value to another asset. This means that any unit of the token can be exchanged for another unit of the same token without any loss of value. Fungible tokens are often used as a medium of exchange, similar to how traditional fiat currencies are used in the real world.

In the context of facility management, fungible tokens could potentially be used to track and manage the usage of resources within a facility, such as electricity or water. For example, a building owner could issue tokens that represent a certain amount of electricity and sell them to tenants. The tenants could then use these tokens to pay for their electricity usage. This could allow for more flexible and efficient billing and payment processes, as well as provide a way for tenants to track their energy usage and potentially incentivize them to use resources more efficiently.

Another potential application of fungible tokens in facility management is in the tracking and management of maintenance and repair work. For example, a building owner could issue tokens that represent a certain amount of maintenance or repair work and sell them to tenants or contractors. The tokens could be used to track the progress of the work and ensure that it is completed in a timely and satisfactory manner.

The use of fungible tokens in the built environment can provide a more transparent and efficient way to track and manage the use of resources and services within our buildings around the world.

Non-fungible Token (NFT)

An NFT, or non-fungible token, is a type of digital asset that represents ownership of a unique digital item or piece of digital content. NFTs are stored on a blockchain and are unique in that they cannot be exchanged for other items of equal value. This is in contrast to fungible assets, which are interchangeable and can be exchanged for other items of equal value, like a security.

NFTs have gained popularity in recent years as a way to authenticate and monetize digital art, music, videos, and other forms of digital media. They can also be used to represent ownership of physical items, such as collectible items, artwork, or even elements (systems) of buildings.

One key feature of NFTs is that they are immutable, meaning that they cannot be modified or altered once they have been created. This ensures that the ownership and authenticity of the asset can be verified and traced. NFTs also allow for the creation of unique digital marketplaces and communities around specific types of content.

We are exploring the use of NFTs in construction documents and facility management documents to identify building systems like electrical, plumbing, and mechanical. By minting these systems as NFTs, we create immutable data that can be authenticated and trusted. With trusted data, the geolocation of building systems can augment as built/record documents to also be trusted, saving time and money. This same data can be used to provide measurement and analytics on the performance of building systems and their materials and equipment.

Smart Contracts

A smart contract is a self-executing contract with the terms of an agreement between buyer and seller being directly written into lines of code. The code and the agreements contained in them are stored and replicated on a blockchain network. Smart contracts allow for the automation of electronic contracts and can be used to facilitate, verify, and enforce the negotiation or performance of a contract. So cool.

Smart contracts can be used in a variety of contexts, such as financial transactions, supply chain management, and real estate. They can help to streamline processes, reduce the risk of fraud, and increase transparency and accountability. Because they are stored on a decentralized, immutable ledger (blockchain), they can also provide a permanent record of the contract and its terms. Overall, smart contracts offer the potential to increase efficiency, reduce the need for intermediaries, and enhance trust in contractual relationships. Smart contracts can be used in facility management to automate various tasks and processes. Here are some examples of how smart contracts could be used in facility management:

- Lease agreements: Smart contracts can be used to automate the process of signing and executing lease agreements for rental properties. The contract can include terms and conditions for the rental, as well as automatic payment and renewal provisions.

- Maintenance and repair: Smart contracts can be used to schedule and track maintenance and repair tasks for a facility. The contract can specify the frequency and scope of the maintenance, as well as the terms for payment to contractors.
- Utility management: Smart contracts can be used to automate the process of managing utilities for a facility, such as electricity, water, and gas. The contract can track usage and automatically bill tenants or occupants for their consumption.
- Security and access control: Smart contracts can be used to manage access to a facility, such as by issuing digital keys or badges to authorized individuals. The contract can also specify the terms under which access is granted or revoked.
- Environmental monitoring: Smart contracts can be used to monitor and track environmental conditions within a facility, such as temperature, humidity, and air quality. The contract can trigger alerts or actions when certain thresholds are exceeded.

When used in construction, smart contracts can be used for:

- Payment Management: Smart contracts can automate the payment process between parties involved in a construction project. The contract can be set up to release payments automatically when certain milestones are reached or specific tasks are completed. This eliminates the need for manual invoicing and reduces the risk of payment disputes.
- Supply Chain Management: Smart contracts can also be used to manage the supply chain in construction projects. The contract can be set up to automatically order materials when inventory reaches a certain level, track shipments, and verify delivery of materials.
- Quality Control: Smart contracts can be used to ensure quality control in construction projects. The contract can be set up to verify that certain standards are met, such as building codes and safety regulations.
- Project Management: Smart contracts can also be used to manage the entire construction project. The contract can be set up to track progress, schedule tasks, and monitor budgets. This can help ensure that the project is completed on time and within budget.

Tokenomics

Tokenomics is a term that refers to the study of the economics of tokens, which are digital assets that can be used to represent ownership, access, or rights within a particular ecosystem or platform. Tokenomics involves the design and analysis of the economic rules and incentives that govern the creation, distribution, and circulation of tokens within a particular system. This can include the issuance and distribution of tokens, the use of tokens as a means of exchange or payment, and the management of the overall supply and demand for tokens. Tokenomics is a complex and multifaceted field that involves economics, psychology, game theory, and computer science, and it is often used to design and analyze decentralized systems and platforms, such as cryptocurrency networks and blockchain-based applications. I find it easy to think of tokens in the context of needing a token to play an arcade game. Once you insert the token, you can play the game.

GPT

GPT (Generative Pre-training Transformer) language model was specifically designed for chatbots and conversation systems. It can be used to build a wide range of conversation-based applications, including facilities management systems. Here are a few examples of how GPT could be used in facility management:

- Building maintenance: GPT can be used to build a chatbot that helps facility managers track and schedule maintenance tasks for various parts of a building, such as elevators, HVAC systems, and electrical systems. The chatbot could allow users to report issues and request maintenance, and then generate responses based on the user's input and the status of the maintenance tasks.
- Meeting room booking: GPT could be used to build a chatbot that helps users book meeting rooms and conference facilities. The chatbot could allow users to search for available rooms based on their location, size, and availability, and then make a booking by providing their name and the dates and times they want to reserve the room.
- Visitor management: GPT could be used to build a chatbot that helps facility managers manage the flow of visitors in a building. The chatbot could allow users to check in and out of the

building, request access to certain areas, and track their movements within the building.

- Energy management: GPT could be used to build a chatbot that helps facility managers monitor and optimize energy usage in a building. The chatbot could allow users to view real-time energy consumption data and suggest ways to reduce energy usage, such as by turning off lights or adjusting thermostats.
- Emergency response: GPT could be used to build a chatbot that helps facility managers handle emergencies in a building. The chatbot could allow users to report emergencies, such as fires or medical emergencies, and provide guidance on what to do in case of an emergency (Figure 1-4).

GPT has the potential to revolutionize the construction industry by automating various tasks and improving efficiency and accuracy. Some examples include:

- Automated documentation: GPT can be used to generate reports and other documents automatically. For example, it can generate reports based on construction progress, project status, and other project-related data.

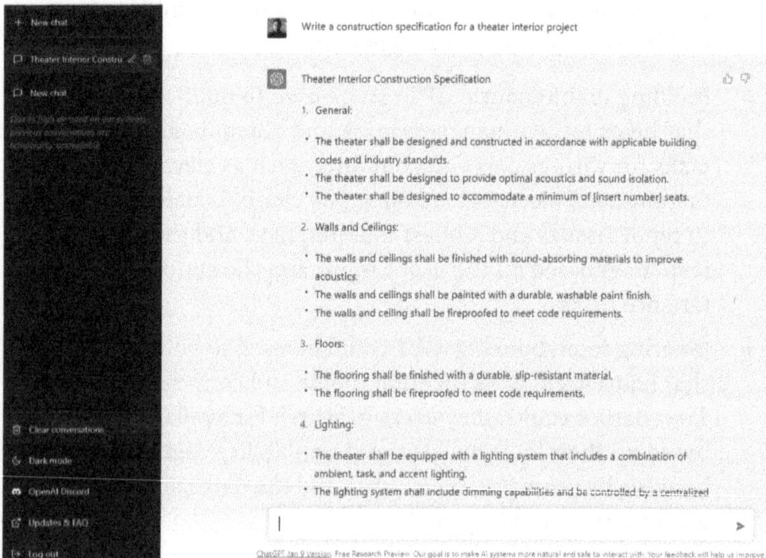

Fig. 1-4: Artificial Intelligence (AI) ChatGPT construction specification example. Courtesy: Paul Doherty.

- Quality control: GPT can be used to analyze and identify errors in construction plans and designs. It can also be used to monitor construction progress and detect potential quality issues.
- Safety monitoring: GPT can be used to monitor construction sites and detect potential safety hazards. It can analyze sensor data such as temperature, humidity, and vibration to identify potential issues.
- Predictive maintenance: GPT can be used to predict maintenance requirements for construction equipment and machinery. It can analyze data such as equipment usage, performance, and environmental conditions to identify potential maintenance needs.
- Cost estimation: GPT can be used to generate cost estimates for construction projects. It can analyze project specifications, materials, and labor requirements to provide accurate cost estimates.

There will be a continuous stream of new terms, phrases, and acronyms for us to learn as we enter and evolve into this decentralized digital world of the metaverse. These definitions are just a starting point for us to begin our journey into the metaverse and will continuously be challenged, changed, and used as our maturation into the metaverse evolves.

CHAPTER 2

Digital Twins, Virtual Worlds, and the Metaverse

As our first deep dive on our journey into the metaverse, let's get beyond the descriptions of the 3D visuals of digital twins, virtual worlds, and the metaverse and into the representations, implementation case studies, and paths forward for our industry. I've purposely put these technology terms of this chapter, digital twins, virtual worlds, and the metaverse, in this specific order to showcase the interrelationships and opportunities that are before us as an industry.

Digital Twins

I went to university at the New York Institute of Technology where I studied architecture. I was offered the opportunity to be part of a Work/Study program where I worked in the real world one semester, then went back to school the next semester. I liked this program as it provided me with internship hours while still I school, shortening my required internship hours to sit for my architectural license exam. I worked for a small architectural firm in New Hyde Park, on Long Island, New York, that provided me with a comprehensive experience of having to do a little bit of everything in order to keep the office going and satisfying clients. I really enjoyed it as I had the opportunity to do things to make the traditional architectural practice better. One of these opportunities was to learn Computer-Assisted Design (CAD) software called AutoCAD by Autodesk. The brother-in-law of the architect I was working for in the Work/Study program ran an exhibit/conferencing business where they would design, build, and pre-stage trade show booths for IBM. Being the late 1980s, IBM was the world's biggest tech company and they exhibited at many trade shows around the US at the time. The architectural firm I worked for shared office space with this exhibit company. The exhibit company needed a designer and they had a lot of IBM computer equipment, so I began to moonlight on my off hours from the architectural firm using state-of-the-art IBM computer equipment running AutoCAD. I was able to design and build on-site IBM exhibit booths for a 4-year period. During that time, I was educated in how computers work, with items like hardware, software, networking, graphic cards, memory cards and how to maximize computing power to showcase software solutions from brand-new tech companies like Microsoft, Adobe, Lotus, and many others. Thus started my love affair with technology, learning it from the inside out and I became very good at it. It was during this time that I began to question the use of technology as only productized as a computer. IBM's tech model for personal computers was to source the components from Other Equipment Manufacturers (OEM), put these components into place according to IBM's design, and then enclose these components in a covered shell with an IBM logo on the outside. To me, it seemed a simple model that could be replicated for physical buildings. As the design and construction industry, couldn't we source computer components, design them into the walls, floors, and ceilings of our buildings and then operate our buildings as computers? And if every building operated as a computer, if we linked these buildings together through technology and created a computer network, have we then created the Internet of Buildings? These thoughts have followed me throughout my career that there should be a seamless method of moving, using, and disseminating digital assets within the structure of physical assets. Interchangeable experiences in a Cyber–Physical environment. Buildings as computers has been my goal, with the emergence of Smart Buildings and Smart Cities becoming a reality recently, I am pleasantly pleased to watch this grow (Figure 2-1).

A big part of this growth of Smart Buildings and Smart Cities will be how to manage them as they are designed, built, and managed. Being an architect, the graphic representation of my designs has

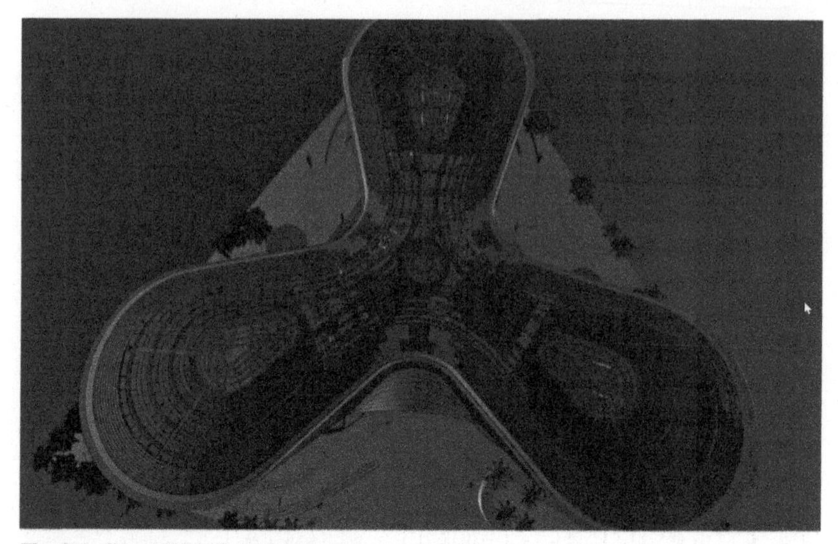

Fig. 2-1: Expo2020 in Dubai Metaverse.
Courtesy: Paul Doherty.

always been a powerful communication method to the builders and contractors of what my intent is regarding the design of my buildings. Having begun my career hand drafting, I have been pushing the profession and the industry to adopt the digital asset of construction documents as I can see its increase in awareness and understanding in digital form better than hand drafting. Beginning with creating a digital version of 2D hand drafting, CAD was a good first step. With the emergence of 3D software, a new method of digital construction documentation was necessary to move beyond drafting/drawing and begin the use of 3D modeling. In the late 1990s, I was introduced to a company in Cambridge, Massachusetts, called Charles River Software. They had developed a 3D solution that they were calling 3D Parametric CAD that allowed the model to update all 2D construction documents automatically if the design changed. This was amazing to me as when changes happen, I used to have to go through each page of the drawing set in order to make sure all the changes were made. With 3D Parametric CAD, with one mouse click, all changes were made saving me an enormous amount of time. I became a consultant to Charles River Software and began to not just use this software but I also introduced this software to my friends who happened to be leaders in influential architecture firms across the US. As a company, we benchmarked ourselves with other emerging 3D CAD companies and noticed that the term Building Information Modeling (BIM) was being used to describe 3D Parametric CAD. BIM was easier to say during a sales pitch, so we began to call our software BIM and changed our company name and software name to Revit. We quickly positioned ourselves as the market leader in BIM and Autodesk took notice. Through a series of negotiations, Autodesk bought Revit and the rest is history with Revit being the de facto standard for BIM in the world today.

BIM becomes an important tool in our industry's adoption of digital twins, virtual worlds and the metaverse. Since we are required to deliver graphic information to construct a building, our industry (and in particular our architectural and engineering designers) already creates the digital DNA of the built environment every day. The challenge our industry faces is that we perceive our CAD documents and Building Information Models to be only useful during the contracted period of time for the design and delivery of a physical asset. Some in our industry are taking the view that CAD/BIM have continued value after the building is delivered in the form of a digital twin that can be used for facility management (FM) and operations, while others are looking beyond FM features and functions and seeing that CAD/BIM can be the building blocks for virtual worlds in the metaverse. Due to the way data is stored in a BIM (The "I" in BIM), many virtual world and metaverse developers who wish to have realistic environments are turning to BIM as a starting point in their visual solutions. In addition, BIM data is also being explored by these developers as a foundation of how a Level of Experience and Level of Engagement (LOE) can enhance a guest's experience. As an example of LOE, my mom is an immigrant from Ireland and has no time for technology except for FaceTime because she can visually communicate with her grandchildren. One Christmas, I purchased an Oculus Quest 2 virtual reality headset for my son. After Christmas dinner, I wanted my mom to try the device. I sat her down in the middle of our living room and had the Quest 2 ready to wear. I loaded the Quest 2 with an application that would bring her to her hometown in Ireland. This particular app uses BIM for some of the scenes and I wanted to see her reaction. She was complaining at a mile a minute that she did not want to "put on that contraption" and she had no use for it at all. As I put the device over her head, she fell quiet. She began to look up and down and around her. She began to giggle and began describing what she was seeing. She could see the road her house was on back in Ireland when she was growing up, the field across the street was now a housing subdivision, and so on. She was happy to experience her

Irish hometown again while sitting in a chair in Memphis, Tennessee. This virtual reality level of experience was measured in a happiness factor. The fact that BIM was used to create stunning visuals begins to measure satisfaction in LOE measures. Her LOE was meant to be visual in nature to elicit an emotion. If we had turned off physics and had her floating above her hometown, she would have felt uncomfortable, so depending on the intended guest experience, a proper LOE of what the technology can do must be implemented. It's all about the levels of experience in the context of what technology to use and what not to use.

Although BIM is a low-hanging fruit in our industry when it comes to digital twins, let's explore the current state of digital twins from a higher altitude so we can develop a pathway to the metaverse.

Digital twins are virtual representations of real space within virtual space, like a digital mirror of reality. But to be clear, CAD or BIM are not digital twins. Digital twins need to move beyond just the graphical information and into a functional state to be valuable. The recent integration of BIM and gaming engines would be an elementary layer of a digital twin that provides movement through the model and assigning simple tasks like identity of an element. As digital twins evolve into digital worlds in the metaverse, a natural order of Levels of Experience/Levels of Engagement (LOE) begins to emerge through the addition of functionality. A formal LOE list is slowly developing as more APIs and data integration into 3D models provide a maturity of BIM becoming a digital twin. One example is the use of Digital Twin Definition Language (DTDL) for BIM. DTDL is based on JSON-LD and is programming-language independent. DTDL can be used to represent device data in other Internet of Things (IoT) services, leading to the intersection of digital twins and Smart Cities.

Digital twins are quickly moving beyond the symbolic 3D geometry of graphic representation and into a relationship with a user's identity, history, and their relationships. When overlaid on what we collectively acknowledge as In Real Life (IRL) an interesting and inspiring new reality begins to emerge.

Industries like aerospace and pharmaceutical manufacturing have been using Digital Twin protocols and processes for decades. The difference today is that there is an application programming interface (API), which is a set of definitions and protocols for building and integrating application software. This allows industry associations, AEC companies, AEC individuals, and the open market to interface together in a digital environment that can incorporate AEC traditional processes of what, why, and how to build the built environment while also capturing and developing the foundation of the digital built environment in the form of virtual worlds in the metaverse.

What we have discovered over time is that digital twins have maturity and usability measures that can assist you in how to implement a digital twin for its most valued use for you. At a basic level, a 3D digital model that represents what is IRL can be a useful digital twin for users that need to identify elements of a building and if connected to technologies, like gaming, provide the ability to virtually walk through the model. This can be helpful for better understanding design decisions, locating elements for operations, and creating the base for other features and functions of the model.

Digital Twins have so much promise today that can report and/ or synchronize to the commands they respond to. This situation provides opportunities for project teams that can describe relationships between digital twins that can provide tremendous value for the immediate needs of a project, but beyond the traditional scope of a project.

Using digital twins, common representation of places, infrastructure, and assets will be paramount for interoperability and enabling data sharing between multiple domains (Figure 2-2).

Fig. 2-2: Expo2020 in Dubai Metaverse.
Courtesy: Paul Doherty.

Gaming

I have a 12-year-old son. He has grown up in the Age of Gaming. I like to think that he and his friends of his age are the natives and we, as the older generations, are the immigrants in the world of digital gaming. I learn from him every day. He began his major push into online digital gaming through the game, Minecraft. He enjoyed playing mostly in creative mode where he could build his digital worlds and learn about the behaviors of objects. He would get lost for hours building and inviting online friends to either join him in the building of these fantasy worlds, or after it was built, to enjoy the experience he created together. He then "graduated" to another digital gaming environment called Roblox. In this game, there are many preconstructed elements and objects that allow my son to quickly design and build virtual worlds without the use of computer code. Roblox has many free elements and objects, but to get a deeper experience and more robust functionality, you can purchase Robux that acts as virtual currency for transactions within Roblox. The ability for users of Roblox to create these virtual worlds and invite others to visit and experience that world has created a large worldwide community that continues to grow every day. Through word of mouth with his friends, he then discovered Fortnite by Epic Games. I find Fortnite to be fascinating on many levels as the producers use many of the same functions and business frameworks of Minecraft and Roblox but deliver a unique and market leading environment whose players are passionate about the game. Like Roblox, Fortnite uses their own digital currency called V-Bucks that is used to purchase a variety of digital assets to enhance your Fortnite game. The favorite digital assets purchased are digital uniforms and costumes for your avatar, as well as weapons and different levels of "superpowers" for your avatar. The avatar is your interface on how you play Fortnite. V-Bucks account for over US$9 billion dollars in annual sales within Fortnite. That is no joke. You can see V-Bucks and Robux hiding in plain sight at any retail or grocery store near the checkout lines, usually a card that is hanging near the discount restaurant gift cards. I use these cards as rewards for a job well done in the physical world or stuffed inside a birthday or Christmas card for my son. It is always appreciated as it enhances his experiences in the digital world.

What I am learning from him about this new digital realm is that his generation does not see a difference between the physical and digital worlds, it is a singular, seamless experience. From chatting walking home from school, to getting into a Fortnite battle online when they get to their individual homes, to when they physically play together where they teach what they have learned with each other (because games like Fortnite have no official "How To" manuals), to sitting around with each other, talking about their latest Fortnite conquests, this generation is seamlessly living in both the physical and digital worlds, and sometimes they enjoy (and prefer) being in the "in between." This "in between" is what blockbuster TV shows like Netflix's *Stranger Things* and legendary motion pictures like *The Matrix* and *Tron* were built upon.

The technology behind all of this is called a gaming engine. Led by both Unity and Epic Games (Unreal Engine), they are providing a gaming engine for our industry's digital assets, like BIM, to be emancipated from traditional use as a construction document, into a new lease on life as a digital asset that has functions, use and value beyond construction documentation.

There are gamers in our industry who grew up with digital games on gaming consoles like Xbox, PlayStation, and others. The new streaming gaming platforms like Twitch and YouTube Gaming are bringing in the masses in a massive way without the expense of a gaming console. The reason this is important for our industry is that we need to recruit talent, and in many cases not traditional talent, in order to provide shelter for the human race. We are beginning to see an increase of the industry's need for gamers that will create new roles and have new responsibilities on a job site and in a designer's studio.

The opportunity to seek change in how we design, build, deliver, and operate our buildings around the world may come from fresh insights in how the digital worlds work in the form of gaming. We must rethink how we create physical environments from learning how people create digital environments. We have a lot to learn about how designers create virtual environments so that we can create a better built environment. It's a backward conversation because usually it's all about the physical asset and we just save as, and we have a digital version. Sometimes the digital version of what's in reality sucks because it's based on something we think the experience should be rather than taking the digital world's view. Gaming is providing a much-needed discussion in our built environment regarding digital twins and the metaverse that needs much more discussion as we navigate the gray waters of digital transformation (Figure 2-3).

Fig. 2-3: Expo2020 in Dubai Metaverse.
Courtesy: Paul Doherty.

Monetization

As you have learned, in the construction industry digital twins can be used to create a virtual representation of a building or infrastructure project, enabling stakeholders to better understand the project's design, construction, and operation. To monetize digital twins in the construction industry, there are several potential revenue streams that companies can explore:

- Consulting services: Companies can offer consulting services to help clients create digital twins of their buildings or infrastructure projects. This can include services such as 3D modeling, data analytics, and simulation.
- Software licensing: Companies can develop and license software that enables clients to create and manage their own digital twins. This can include features such as data visualization, simulation, and predictive analytics.
- Maintenance and support: Companies can offer ongoing maintenance and support services for digital twins, including software updates, bug fixes, and technical support.
- Data analytics: Companies can offer data analytics services that use data from digital twins to provide insights into building performance and help clients optimize their operations.
- Advertising and sponsorship: Companies can monetize digital twins by offering advertising and sponsorship opportunities

within the virtual environment. For example, a building materials company could sponsor a virtual trade show hosted within a digital twin of a construction project.

Another monetization use case for a Digital Twin is as a Digital Birth Certificate for a physical asset. In 2009, I became the president of a 3D digital real estate marketing company called Screampoint. Screampoint created beautiful 3D digital assets for large-scale real estate developers around the world. We were constantly asked what else our photorealistic 3D models could do beyond showcasing a building before it was built. We often worked with the marketing and sales teams of these developers who wanted to use our 3D models for things like sales kiosk integration, broadcast advertisements, and simulations. What we found over time is that our 3D digital assets were so integrated into the physical assets that when time came for the sale of the physical asset, part of the sale valuation was contingent that the 3D digital model and subsequent uses of that model for the building increased the sales price and was written into the deed of the building. It was like we created a Digital Birth Certificate for that building. Dozens of property sales have used the Screampoint models as part of their property asset value during a real estate transaction. From our own experience, I feel that we are entering an era where the digital representation of physical real estate assets will become part of the overall property valuation and transaction in all real estate transactions. A Cyber–Physical relationship built on money will turn some heads. The key to monetizing digital twins in the construction industry is to create value for clients by helping them optimize their operations, reduce costs, and improve the quality of their projects.

A good resource for further exploration of digital twins can be found in a report called, *Digital Twins from Design to Handover of Constructed Assets*, Published by the Royal Institution of Chartered Surveyors (RICS), March 2022 https://www.rics.org/news-insights/research-and-insights/digital-twins-from-design-to-handover-of-constructed-assets.

Virtual Worlds

Virtual worlds seem to be the proper definition of how the world creates 3D web sites and then begins a process of linking them together like hyperlinks in the Web 2 world. I like to describe and think of virtual worlds that operate according to their own laws, and the passage of time is no exception. It all speaks to concepts beyond the experience of time's passage. In virtual worlds in the metaverse, the future supersedes the present, just as what might be guides vision to a greater extent than what is in reality.

Virtual worlds can be derived from scanning physical spaces (reality capture) or constructed via artistic or other design/ worldbuilding activities like CAD and BIM. These virtual worlds in the metaverse allow people to look remotely through the eyes of other people to gain their perspectives of what they see physically. There are unlimited applications that can be implemented via the messages that are delivered by the medium of virtual worlds in the metaverse. The most exciting aspect of virtual worlds in the metaverse is the emergence of these new applications of how it can be used. The great hope is that the metaverse can unlock human intelligence and link minds so as to address globally the pervasive problems of our civilization.[1]

To create virtual worlds, there are a number of processes that you can use to expand your journey into the metaverse. We've discussed BIM earlier in this book, but one of the more popular methods of creating a virtual world is to use Reality Capture and Motion Capture to digitally represent what is in the physical world.

Reality Capture and Motion Capture

In the physical world, we acquire spatial information with our eyes and build a 3D reconstitution of the world in our brain, where we know the exact location of each object. Similarly, virtual worlds in the metaverse need to acquire the 3D structure of an unknown environment and sense its locations, scale, and motion.

Reality capture of the built environment refers to the process of creating a digital representation or 3D model of physical spaces and objects in the real world. This is typically done using various technologies such as laser scanning, photogrammetry, and structured light scanning. The captured data is then processed and converted into a highly accurate and detailed model that can be used for various applications, such as architecture, construction, engineering, and urban planning. The goal of reality capture is to provide an accurate and realistic representation of the built environment that can be used to improve design, construction, and maintenance processes.

---

[1] Thomas A. Furness III. Professor, University of Washington, www.versemaker.org.

Motion capture, also known as mocap, is a technology used to capture the movement of objects or people in real-time, often for the purpose of creating realistic animations or simulations. In the built environment, motion capture can be used to track the movement of people or vehicles within a space, allowing architects, designers, and engineers to analyze and optimize the design of buildings, streets, and other urban environments. This can include analyzing pedestrian traffic patterns in public spaces, optimizing the layout of parking lots and transportation hubs, or studying the movement of vehicles on highways and roads.

Chase Olson – Reality Capture

I thought it would be good in this chapter to interview a world-class expert in the reality capture space. I am lucky to know and be friends with Chase Olson of Smart Sky Tech Hub, one of the world's leading reality capture companies. In the following interview, Chase provides insight into what reality capture is, how to use the captured reality, and where he sees this going.

CHASE: I'm Chase Olson. I'm originally North American. I was born in Wyoming and raised in Salt Lake City, Utah. However, I had the opportunity after finishing my degree in International Business to come down to Brazil, as an entrepreneur been here since 2012. And as of about 2014, I really started getting into the reality capture world; right back then, we were using the term phygital (Physical–Digital) rather than metaverse. I've been working for the last five years with a Brazilian startup called Smart Sky Tech Hub. The idea is working with digital spaces, however, 100% of the time, focused on a physical asset or a physical point of reference, so that we're not just creating new worlds. And the reason for that is we feel like we have so much information at our fingertips. I mean, when we use platforms like (ESRI) ArcGIS, we're able to understand a city. And all that information is geo referencing is accessible. However, it's very limited. And in our view of what the future of the metaverse what's the future of Web3, and how they connect together to create interactive and physical experiences, is working with reality capture technologies to bring a physical asset to a digital realm. I don't know about you, Paul. But I've been in the industry for about five years. And I know, maybe less than 10% of people in the industry have access to a good set of VR glasses at home. So I question, what's the future of virtual reality if the people in the field don't even use it on their day-to-day basis? If it's 100%, virtual, and that's where we come from, it's augmented, it's virtual, it's interacting in a digital world.

PAUL: What do you mean by reality capture?

CHASE: Reality capture (Recap), believe it or not, is really just what it's saying. It's capturing reality. And what we mean by that is the use of technologies like drones, LiDAR (which is laser scanning), camera 360, pretty much any sort of sensor that's able to go into the field, create a digital footprint of what was captured via datasets, which are called Point Clouds, which are essentially just millions of pixels. Each pixel has four characteristics, longitude, latitude, altitude, and color. And with that, you're able to replicate any sort of physical asset. We use existing BIM, like (Autodesk) Revit, for anything that has to do with the building project itself. Using recap, we can capture existing buildings, bridges, shopping centers, anything that's the asset itself, we would all do in the field. Anything in the infrastructure side of things (streets, the transmission lines, the light poles) we 3D model using (Autodesk) Infraworks. We then consolidate all that into one unified model. From there, you're able to define the information and create families within a model. A wall, a ceiling, a door, are examples of families. Inside of the family, you have all of the elements that are put together to create that family. So inside of a wall, you might have concrete, or you might have wood, and you might have all of these different materials, as well as an understanding of how many man hours it might take to create that wall and things of that nature. Once we add all this type of information, then we are truly working with a digital twin. We find that going from a digital twin to a Metaverse experience is really just plug and play. It's just understanding extensions, understanding file types, how to convert them. And you can upload any digital twin into an Unreal Engine, or a Unity engine easily. How you access the digital twin, that's where a lot of our specialty comes in. That's essentially the recap process.

Recap is being used in many phases of a project. For instance, we have a lot of construction companies looking at reality capture trying to bring it into their workflows in the same way they work with traditional surveys or traditional analysis of some sort. And it's very inefficient. Because the cost per reality capture is the majority of the time a little bit more expensive than a traditional, but it's much more efficient, much quicker. But when you use it for one phase of the job, it might not be the most cost-efficient process. However, with reality capture different to a traditional, you didn't just generate the data that you needed for that moment, you generated all the possible data that was available at that moment, which means that when you start understanding that the data is multidisciplinary, you're able to use it not just in engineering, you can use it in your projection phase your budgeting, you can use it in your control your quality control, you can obviously use it in marketing and commercial as a subsequence. You can use it to improve the security of the jobsite, when you start using it in a multifaceted way, all of the different departments within the company, that's when you have a lot of gain and the value of recap becomes clear. And that's where we focus our time in helping our owners or our clients understand via our portfolio of products, that now from one investment in the field, we're able to generate we've already generated up to I think the most was 18 products off of one single captor rather than one to one which is the traditional way of working.

PAUL: This consolidated model is super interesting as you are getting closer to identifying and managing the trusted data. This is a big deal because you have to trust the data in order to use it properly (Figure 2-4).

CHASE: Let me provide an example. Let's say I'm design-building a greenfield housing development. Traditionally, it can take a long time to study the site, provide diligence and a feasibility report. I'm going to invest initially in a survey study, and then I'm going to do perform my studies, analyses, and reports from that, I'm going to have to take those projects to the city and get aligned with the city to get permission to build, I'm going to have to take those to my investors or to the bank as well for my financials, and everyone has to be

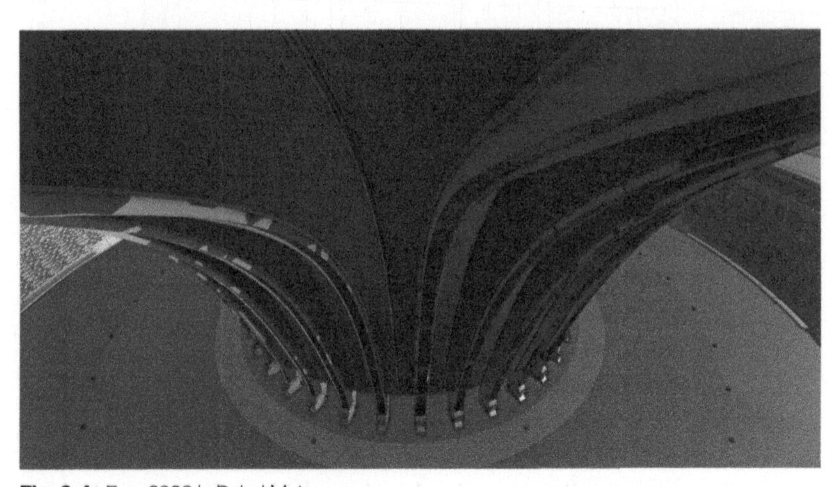

Fig. 2-4: Expo2020 in Dubai Metaverse.
Courtesy: Paul Doherty.

aligned on this. What often happens during that phase is that I'm unable to execute my project the way that I thought I was going to governance, environmental, and financial issues. What also happens is we end up altering the project several times during the phase, but the area where the project is going to be developed hasn't changed at all. But since there are changes and revisions to the master plan design, for each change, a developer has the survey team go back out and redo work. With recap, the traditional process changes to save time and money.

The recap process has already developed what's called geodesic. We create a field with geodesic reference points (ground control points) which are then fixed into the ground. Geodesics are strategically positioned in areas of the job that will have the least amount of volume or any sort of movement. And then all we have to do is execute the drone or other reality capture technology and not have to engage the entire topography process. By setting the baseline as geodesic information, the consolidated model is the foundation for the entire development throughout the entire project.

PAUL: The augmentation of traditional construction documents through recap data is enriching the source of truth for that project, no matter if it's horizontal or vertical, which is quite remarkable. So, what I've learned is that the consolidate model of the project can be sliced and diced and used for a specific role at a specific time in the lifecycle of that physical asset. It can be used for pre-design, pre-construction, design, construction, facility management, operations, and maintenance. But an interesting shift in thinking is that the built environment can be replicated and can become a digital twin that could morph into a virtual world using the consolidated model with gaming engine technology.

This lifecycle approach to using the recap consolidate model forces a conversation of how does a model of truth of reality also become the model of truth for unreality, like virtual worlds in the metaverse?

CHASE: That's where it's a very difficult subject. As of now, the regulations in this field of work are almost non-existent. Even in the world of gaming, there is very little energy or effort in creating an opensource, multi-platform multiverse where my avatar made in Fortnite can play in a League of Legends game. You have Facebook (Meta), you have Microsoft, you have Apple, they're all trying to create the metaverse. First off, I don't think that's even not even really like a possible thing to happen. You can't really own it in that way. Because for you to own it, you'd have to then integrate with all the other competitors, or else it's not a Metaverse. I really liked Matthew Balls' book, *The Metaverse and How it will Revolutionize Everything*, where he calls those company's efforts "meta galaxies." I love that term. The fact that by capturing the reality of the built world, our company has a lot of responsibility to maintain this data and protect this data. We plan on never being acquired, especially by one of the big companies. We have a lot of very powerful information that in the hands of one of the big players could be potentially dangerous.

PAUL: This begs a huge question. Being a service provider and having the responsibility of being the steward of the built environment data, that digital DNA about our physical world, as these demands of the digital realm start to intersect with the metaverse and major tech players, do you foresee in your crystal ball that there will be, or needs, some form of regulation, because this isn't about posting up funny pictures up on Instagram? This is some serious information that in the hands of bad actors could be used against us. Do you foresee something like that?

CHASE: I do think it's always a professional responsibility. But you can't really count on that when you're talking about a global activity. I think that it is something that we are going to have to create. And I don't think that's possible without these of entities and agencies that do this type of monitoring and auditing without them evolving with the use of innovations like blockchain technologies. I speak a lot about digital twins and the metaverse. But when I go out and present the entire workflow, I speak about blockchain, tokens, NFT's, DAOs, Web3, and I then finalized with the metaverse, because for me, out of all of those, the metaverse is nothing more than how we interact with everything else.

PAUL: Thank you for this insight, Chase.

Avatars

One of the coolest but most challenging aspects of gaming in the built environment is having a cyber–physical experience as an avatar. Avatars in virtual worlds in the metaverse are digital representations of users/players that allow them to interact with the virtual environment and other users in a more immersive way. These avatars can be customized to reflect the user's preferences, identity, and personality. In virtual worlds in the metaverse, avatars can move, speak, and interact with other objects and users. They can also have unique abilities, such as flying or teleportation, depending on the rules of the virtual world they inhabit. Avatars can be designed to look like realistic human beings, cartoon characters, animals, or anything in between. Avatars play a crucial role in the metaverse because they allow users to create a virtual identity and engage with the virtual environment in a more meaningful way. Users can use their avatars to attend events, socialize with other users, play games, or explore the virtual world. Avatars can also be used to represent businesses or organizations, allowing them to establish a virtual presence and engage with their customers in a more immersive way. Avatars are a very important element when designing your presence for marketing/sales purposes and for when you are on "virtual site" for a built environment project (Figure 2-5).

Fig. 2-5: A Steampunck Avatar in the Metaverse.
Courtesy: Paul Doherty.

Avatar Interface

Epic Games has a business unit that creates metahuman avatars that capture motion-capturing movements when a human smiles and gets dimples or all the little idiosyncrasies that person has. And that Metahuman remembers this physical action over time. What do her eyes look like when they smile? What does their language look like they talk? These new ways of communicating are now all around us. They will integrate seamlessly without a language barrier especially with my son's generation. In our contemporary climate, where physiological gender is itself in a state of flux, mere aesthetic representation of the user is far less important than the authenticity of interactions and relationships built in the metaverse space.[2]

[2] The Metaverse as Virtual Heterotopia, David van der Merwe, 2021, www.socialsciencesconf.org.

Avatars as Metahumans/ Humanoids (MoCap)

Taking into account that gaming is now part of the built environments technology stack, we should be able to think of gaming in all its realities to gain insight and the value of gaming for our industry.

A first approach could be an ethical design of avatars and their corresponding behaviors/representation in cyberspace which is a complicated issue. The metaverse could create a gray area for propagating offensive messages, e.g. race and could raise debate and prompt a new perspective to our identity. Could you be an avatar that would experience and feel bias, racism, and abuse to raise your own consciousness for others IRL lives?

If an avatar creates a new identity of oneself in the metaverse, an avatar potentially raises a debate and prompts new thinking of human life. That is, the digital clone of humanity in the metaverse will live forever. Thus, even if the physical body, in reality, is annihilated, you in the digital world will continue to live in the metauniverse, retaining your personality, behavioral logic, and even memories in the real world. If this is the case, the metaverse avatars bring technical and design issues and ethical issues of the digital self. Is the long-lasting avatar able to fulfil human rights and obligations? Can it inherit my property? Is the husband and father of a child IRL still "alive" and what rights does the avatar retain? I find it amazing that in our lifetime, we are required to have these conversations. But here we are.

An interesting scenario that we are discussing as a company for a mega project we are developing provides the opportunity to use our construction site as a never-ending game. A person or company could create a number of avatars as employees to assist IRL workers and conduct portions of work while the IRL workers rest at night. With the emergence of deep fake avatars that leverage AI, construction sites could become digital gaming sites for the Industrial Metaverse taking into consideration the personal rights and responsibilities of the physical and digital workers on-site (Figure 2-6).

NPCs as Reference Oracles

A Non-player Character (NPC) refers to a character not controlled by a player in scenes in a digital video game. The history of NPCs in games could be traced back to arcade games, in which the mobility patterns of enemies will be more and more complex along with each

Fig. 2-6: Zulu Nation–inspired architecture being built in South Africa, but available now in the Metaverse.
Courtesy: Paul Doherty.

level increasing. With the increasing requirements for realism in video games, AI is applied to NPCs to mimic the intelligent behavior of players to meet players' expectations on entertainment. The intelligence of NPCs is reflected in multiple aspects, including control strategy, realistic character animations, fantastic graphics, and voice. In our company, we see NPCs as important elements to an Industrial Metaverse as they can act as expertise reference players. In one instance, an NPC can be the reference for a specialty trade on how work gets done. On-the-job training, situational education, and best practice information for implementation for that specialty trade are all available at any time through the NPC. Non-player characters (NPCs) on a construction site in the metaverse could serve a variety of use cases, including:

- Simulation: NPCs could be used to simulate real-life construction scenarios and train workers in a virtual environment. This would allow workers to gain experience and develop skills without the risk of injury or damage to property.

- Safety training: NPCs could be used to simulate dangerous scenarios and teach workers how to respond to emergencies, such as fires, accidents, and structural collapses. This would improve safety on construction sites and reduce the risk of injuries.
- Quality control: NPCs could be used to monitor construction activities and check for quality control issues, such as deviations from plans, poor workmanship, or material defects. This would ensure that construction projects are completed to the highest standards and minimize the need for rework or corrections.
- Resource allocation: NPCs could be used to manage resources, such as labor, equipment, and materials, on construction sites. They could assign tasks to workers, schedule deliveries, and track inventory levels to ensure that construction projects are completed on time and within budget.
- Project management: NPCs could be used to assist project managers in planning and executing construction projects. They could provide real-time feedback on progress, identify bottlenecks and risks, and suggest ways to optimize construction processes.

Virtual Worlds in the Metaverse Examples

Minecraft https://www.minecraft.net/en-us: Minecraft started as a simple computer game that was built on mining and crafting virtual worlds with Lego-like blocks. Initially there were two main modes: survival and creative. In survival mode, players found supplies and foods to craft tools and avoid creatures. In creative mode, players get given supplies to build whatever they want, they can fly, and do not need to virtually eat. Users can explore pre-existing worlds created by other players and play games within the world such as Cops and Robbers. Minecraft has 180 million monthly active users in 2023.

Roblox https://www.roblox.com/: Roblox is a free gaming platform, where users download a game to play from a wide range of Roblox games. Some of the most popular games on Roblox include Adopt Me!, where users care for virtual pets, which can be traded with other players. Jailbreak is an example of another game where users play-in a virtual "cops and robbers" style game. And Brookhaven RP is a game where users get jobs, purchase houses and cars, and hang out – a virtual everyday life experience. Roblox provides the environment and tools to its community to create their own games. There are millions of different games in Roblox and while they all exist within the same graphical environment they are all very different. Users can download every game in the environment without any additional software (unless you wish to create a game). In 2023, Roblox has 215 million monthly active users.

Fortnite https://www.fortnite.com/?lang=en-US: Fortnite is a videogame where players drop onto an island and battle to be the last player standing. Players can create an island of their own, with their own rules or can play other game modes such as "Save the World." Fortnite's metaverse provides an interactive social network in 3D, where players pick an avatar and explore various islands. Fortnite has collaborated with big corporations and brands for avatar "skins," e.g. John Wick, NFL players, Batman and Catwoman, and Kylo Ren from *Star Wars*. Epic Games' (the company which owns Fortnite) CEO has described Fortnite's user experience as a metaverse because it's not just a simple game, but rather it's a shared experience that allows players to explore the virtual world. Fortnite has 250 million monthly active users, though Epic Games has a registered user count of over 400 million in 2023.

Niantic https://nianticlabs.com/?hl=en: Niantic created Pokémon Go (a smartphone app) via its Lightship platform. Based on augmented reality (AR) technology, it allows users to interact with digital objects in the physical world. The Lightship platform is an ARDK (Augmented Reality Development Kit) designed to help developers build experiences using the same building blocks Niantic used for Pokémon Go. Niantic's long-term goal is to build a 3D map of the world. It is working with Qualcomm to design AR glasses that orient themselves using Niantic's maps, in order to merge the virtual and physical worlds. The app has 80 million monthly users with 572 million cumulative downloads.

Decentraland https://decentraland.org/: Decentraland is a decentralized, immersive virtual world where users interact with others in real time, engage in virtual experiences, and have the ability to trade with one another. Decentraland is another metaverse on the Ethereum blockchain, using smart contracts rather than a centralized entity to verify transactions. Users are able to monetize their virtual creations and experience, just as in the real world. Users can purchase plots of virtual "LAND" and build virtual houses or set up virtual businesses, in addition to playing games within the metaverse. Users who own MANA (the currency of the metaverse) can vote on platform policies and the world is run by its users. Decentraland was officially launched in February 2020. It is overseen by the non-profit Decentraland Foundation (independent of the founders) and the future of the metaverse is decided by the Decentraland DAO (Decentralized Autonomous Organisation). Decentraland had 240,000 monthly active users in Q3 2022. Users can access Decentraland through a web browser without specific software requirements. Decentraland is not currently on mobile devices but can be run on multiple computers as long as the user's digital wallet is installed on each device.

The Sandbox https://www.sandbox.game/en/: The Sandbox is an Ethereum-based metaverse which allows users to purchase virtual land (NFTs) as well as play games and enjoy virtual experiences. The Sandbox aims to help developers monetize content on the Ethereum blockchain, e.g. virtual art, and other assets to help users generate a virtual income, which is transferable into real-world cash. The Sandbox DAO enables users to vote on Foundation grants and other game parameters. Similar to Decentraland, the DAO is governed by SAND token holders and asset owners. The Sandbox metaverse has more than two million registered users as of March 2022. SAND is the in-game currency and LAND refers to digital real estate; each LAND is an NFT on the Ethereum blockchain. There are 166,464 LANDS. Users generate revenue by creating ASSETS, owning LAND, and renting or selling it, or building games and experiencing using LAND.

Engage VR https://engagevr.io/: Engage VR works with commercial clients, to develop virtual worlds. Use cases include employee onboarding, training, product demos, wellbeing, customer outreach, and professional events through metaverses. Engage VR can be accessed through a range of devices including VR headsets, PCs, smartphones (iOS and Android), and tablets. Engage also offers MetaWorlds (i.e. a virtual world within their Metaverse). These worlds can be used as templates, as well as having the option to design something from scratch. These virtual worlds can be shared privately with an organization, class, or small group and can be used for virtual games and immersive activities (e.g. for team-building exercises). The Engage network has over 300 MetaWorlds currently on its platform. The company is also building something it calls the Engage Oasis that is a Metaverse designed for professionals which links MetaWorlds via portals and bridges in publicly accessible plazas. Each plaza has its own theme, and clients with a public Metaworld can choose to list it via a portal inside the plaza structure. The initial plazas at launch will be enterprise, education, and creative. The company CEO David Whelan has described Engage Oasis as the LinkedIn of the virtual world. Engage VR has 160+ commercial clients from blue-chip companies, enterprise clients, and educational clients.

Horizon Worlds https://www.meta.com/horizon-worlds/: Horizon Worlds is Meta's social virtual reality app and it has a number of ready-made VR worlds. It also has the option for users to create their own worlds too. Users can play games, socialize with friends, or even attend virtual events. Just like on Facebook, users can add friends and send messages via the Oculus app. There is a party system that enables users to travel between worlds and communicate with people in their group. The platform also lets users take photos within the app, meaning users can take virtual selfies, essential for 21st century socializing. The app allows users to create mini-games and activities on top of the base game. Users over 18 can create a floating (legless) avatar and explore worlds within the app (e.g. shooting games, river cruises, magic flying broomstick world, and platforming games). Between December 2021 and February 2022 Horizon Worlds grew 10× to 300,000 users.[3]

Cybersecurity and Safety

As virtual worlds in the metaverse continue to grow and become more integrated into our daily lives, it's important to consider cybersecurity best practices to protect yourself and your assets in this digital space. A foundation of best practice for cybersecurity is to use strong and unique passwords. As with any online account, it's important to use strong and unique passwords for all of your virtual worlds and metaverse accounts. You should avoid using the same password across multiple accounts, as this can make it easier for bad actors to gain access to multiple accounts if one password is compromised. You should also enable two-factor authentication. Two-factor authentication (2FA) adds an extra layer of security to your accounts by requiring a second form of verification, such as a code sent to your phone, in addition to your password. This can help prevent unauthorized access to your accounts. Be cautious of phishing scams. Phishing scams can be a common way for hackers to gain access to your accounts. Be wary of any suspicious emails or messages asking for

[3] The Metaverse Multiverse, Research Department of HSBC, March 2022.

personal information, and only click on links from trusted sources. Another best practice is to keep your software up to date. Make sure to regularly update your software, including any virtual world and metaverse-specific software or applications. These updates often include security patches to address known vulnerabilities. Limiting the personal information you share in a virtual world or metaverse setting is important. Be cautious about sharing personal information, such as your real name, address, or financial information. Only share information when necessary, and make sure you're sharing it with a trusted source. In our company, we require that all employees use a virtual private network (VPN): A VPN can help protect your online activity by encrypting your internet connection and masking your IP address. This can help prevent hackers from intercepting your data or tracking your online activity. And as a final security warning, be aware of scams and fraud by being cautious of any offers that seem too good to be true and avoid sending money or virtual assets to unknown or unverified sources. Scams and fraud can be common in virtual worlds and the metaverse, so it's important to be vigilant and use common sense.

Metaverse

The Metaverse is emerging, but it is not here yet. Some are even calling the metaverse the 3D Internet. Like the emergence of the World Wide Web in the 90s, there will "canned" versions of the 3D Internet (AOL, Compuserve, Prodigy) that will emerge with Search for the ability to navigate (Yahoo, Alta Vista, "Portals") that then emerge into larger tasks and features beyond Search (Google, Facebook, Amazon, etc.). The 3D Internet will follow a similar path. For instance, Meta's Horizon as an example of a "canned" version of the 3D Internet, while solutions like ChatGPT could be our first glimpse of Search in the 3D Internet.

The metaverse that is being created has the opportunity to assist our industry in the following ways: (i) increase accuracy and confidence in design and construction documentation; (ii) increase trust and authenticity in the digital data meant for facility management and operations of buildings and infrastructure; (iii) provide a construct of a bi-directional communication and relationship between the physical world and its digital equivalent; and (iv) deliver new and ever-evolving environments for experiences for people who live, work, play, and learn. "Meta" is an ancient Greek word that means "beyond" in a self-referring sense, e.g. a meta-conversation is a conversation about conversations, and metadata is data about data. This means that the metaverse is a universe of universes that provides the ultimate sensory experience. The industrial metaverse and the social metaverse will have profound impacts on our daily work, play, and life, across all industries and sectors, and on the economy while potentially reshaping society for all humankind.

Metaverse and Sustainability are complementary to each other and potentially can be the two mega themes in the coming decades. The Sustainability of the Metaverse will require a concern not only for the Environment (that which surrounds us) but also for the "Invironment" (ourselves, i.e. the human element). In this sense the Metaverse is inextricably intertwined with the Vironment (the interplay and boundary between humans and their surroundings), i.e. space in which humans can interact with computer-generated content and with other humans.[4]

The immediate connection between a cyberspace of virtual worlds unregulated by recognized human governmental authority but still maintained by a strict protocol of rules established between users and owners is still a dream but may have the beginning foundation of becoming reality due to past history. I wrote a book in 1997 called, *Cyberplaces: The Internet Guide for Architects, Engineers, Contractors, and Facility Managers*, published by R.S. Means. In this publication, I wrote in the last chapter about a project that needed a technology innovation called Virtual Reality Markup Language (VRML) to provide a solution to showcase a 3D model on a mobile device of the new San Francisco Giants baseball team's new stadium. Instead of purchasing an engineering software license to show a large number of stakeholders the design of the new baseball stadium, we wanted to show the stadium in a VRML environment of a web browser. Our team, led by Planet 9, was able to deliver a 3D model of the new San Francisco Giants baseball stadium in a web browser that was published in my book as an example of the Metaverse in action in 1997. I wrote that the near-obsessive, practically religious fervor with which the user's interface with these environments makes them more a cyberplace than cyberspace, thus the name of the book (Figure 2-7).

The metaverse is something of a paradox, both an inhabited and enacted space as well as an impermanent interface between the digital

[4] Metaverse Landscape & Outlook: Metaverse Decoded by Top Experts, Matthew Ball, 2022.

Fig. 2-7: A metaverse party with multiple people who are able to interact in real time as avatars. Courtesy: Amit Chopra of Iconic Engine.

and physical realms. But within this paradox, there is the opportunity to seamlessly weave between the physical and digital worlds. As we design our buildings for our real estate developments, we challenge our designers to think of our buildings not just as computers, but as portals to another world. We envision that physical buildings have cyberphysical portals to bring you from one dimension to another seamlessly. This will provide unrestricted expansion across the metaverse, transporting you as an observer or immersively as an actor (as an avatar) into experiences that blur the differences of physical and digital.

What the metaverse proposes, like so many technologically driven transformative processes, is a re-evaluation of the internet itself. Until now, the cyberspace we have interacted with has been localized, stored at first within physical servers and more recently, the vague and nebulous space called the "cloud." As it grows and develops, the metaverse will evolve and iterate, shifting the balance until we, as its users, "will constantly be 'within' the internet, rather than have access to it, and within the billions of interconnected computers around us" (Ball, 2021). Once the digital dust clears the same argument will still hold true: IRL as we know and experience it will still exist, only no longer as an external gateway to the metaverse, but a physical extension thereof.[5]

[5] The Metaverse as Virtual Heterotopia, David van der Merwe, 2021, www.socialsciencesconf.org.

The metaverse allows participants to occupy common virtual 3D spaces (e.g. virtual worlds) from anywhere in the physical world, wherein they can learn, train, design, conduct business, receive diagnosis, therapy, and engage in entertainment activities. The metaverse overcomes the tyranny of distance in allowing participants to be in a place and interact in that space with others without moving their bodies to that place. The incorporation of immersive virtual worlds effectively activates spatial memory that in turn prolongs retention of experiences and what has been learned. This brings a participants' bias of what they have experienced IRL to the digital environment of the metaverse. In a virtual world in the metaverse, we know that we do not need roofs or other overhead structures in our virtual buildings because there is nothing to protect a person from things like rain, snow, wind, or falling objects. But due to people's memory that a building must have overhead enclosures, we need to provide this visual element into our digital worlds although it is not necessary. It provides a level of comfort for people in our virtual worlds. This Level of Experience/Level of Engagement will continue to be a recurring theme as we mature with our designing and building of the metaverse. The fact that we will need to "fake" elements in the metaverse in order to maintain a person's memory of how physics and natural laws work will be a design method that will take generations to overcome. For instance, creating virtual water will need to have every molecule of H20 conform to the laws of nature in the physical world, although virtual water does not need to conform in the digital world.

Industrial Metaverse

The industrial metaverse refers to a virtual space where businesses and industries can operate in a highly immersive and interconnected way. It is essentially an extension of the concept of the metaverse, which is a collective virtual shared space where people can interact with each other digitally in a fully immersive environment.

In the context of the industrial metaverse, companies can use virtual reality and augmented reality technologies to create immersive simulations of their products and services and allow customers to experience them in a highly engaging and interactive way. This can enable businesses to showcase their products in a more realistic way, reduce the need for physical prototyping and testing, and accelerate the design and development process.

The industrial metaverse can also facilitate remote collaboration and communication between employees, partners, and customers across different geographies, which can lead to increased productivity and efficiency. Furthermore, it can enable businesses to collect and analyze large amounts of data in real time, which can inform decision-making and improve operational performance. It involves the use of advanced technologies such as augmented reality, virtual reality, IoT, and artificial intelligence to enhance industrial processes and outcomes.

Here are some examples of how the industrial metaverse can be used in the construction industry:

- BIM Collaboration: Building Information Modeling (BIM) is a powerful tool for construction project planning and management. With the help of the industrial metaverse, BIM models can be shared and collaborated on in real-time across various locations and devices. This can improve coordination between project teams and reduce errors and rework.
- Virtual Site Inspection: Virtual site inspections using augmented reality can be conducted before construction begins. This can help identify potential issues and hazards and improve safety on the job site. The virtual inspection can also be shared with stakeholders to get their feedback and approvals.
- Remote Training and Education: The industrial metaverse can be used to provide remote training and education to construction workers. This can include virtual simulations and interactive learning modules that allow workers to learn new skills and techniques in a safe and controlled environment.
- Predictive Maintenance: The use of IoT sensors and artificial intelligence can help predict when equipment and machinery may need maintenance or repairs. This can prevent costly downtime and improve overall efficiency.
- Digital Twin: A digital twin is a virtual replica of a physical asset or system. By creating a digital twin of a building or infrastructure project, stakeholders can test different scenarios and make better-informed decisions. This can help reduce risks and optimize resources.

The industrial metaverse has the potential to transform the way businesses operate and interact with their customers, partners, and employees, and drive innovation and growth in our industry. As time moves forward, I will enjoy how both the industrial metaverse and the larger social metaverse continue their maturity and growth. I am hoping as the metaverse grows that we always consider what Tony Parisi defined as the 7 rules of the metaverse:

Rule #1. There is only one Metaverse

Rule #2: The Metaverse is for Everyone

Rule #3: Nobody Controls the Metaverse

Rule #4: The Metaverse is Open

Rule #5: The Metaverse is Hardware-Independent

Rule #6: The Metaverse is a Network

Rule #7: The Metaverse is the Internet

I am looking forward to seeing how the talented people in our industry adapt and adopt these technology tools to transform ourselves from traditional means and methods into inspiring solutions to collectively build and sustain our built environment for a better future for all.

CHAPTER 3

Metaverse Mechanisms and Solutions

Although digital twins, virtual worlds and the metaverse, are the 3D Internet's visual elements, the real power comes in the new mechanisms of how to create features and functions. In this chapter we are going to explore the use of blockchain, smart contracts, digital real estate, Web3, and AI innovations that can be seen as the building blocks to a truly immersive, inclusive, and diverse metaverse.

A basic goal of the metaverse is to build the next wave computing platform that houses several sub-platforms accessible through the same 3D interface. The vision is to provide a seamless experience across virtual and augmented realities that can deepen human connections using an ecosystem of service experiences regardless of physical distance. So, while this chapter will cover a seemingly disparate collection of digital innovations, when put together in workflows and use cases, these innovations cause results that only a few years ago seemed like magic. The mechanisms of the metaverse provide new systems, procedures, and techniques that are quickly advancing the maturity of virtual worlds in the metaverse.

The velocity of this advancement of the maturation cycle provides a need for our industry to discuss an important question. The use and adoption of the metaverse in the built environment will only be successful if accessibility to metaverse technologies is open to all. If the metaverse does not allow for 100% access, especially the non-tech savvy, disabled persons, or economically challenged people, the metaverse becomes yet another tool that widens the inequality gap that is already deep rooted in our industry. Since our industry works as an eco-system in order to design, construct, and manage the built environment, then we must have 100% inclusivity in order for the metaverse to provide benefit and value. No one left behind. Perhaps this book can be one of many tools that educates, inspires, and guides all people in our industry to ensure that the metaverse fulfills its promise. Let's explore the metaverse mechanisms and how they apply to you.

Blockchain

At its inception, Blockchain was developed as an effort to take back ownership of the internet and return control of the online world to its users rather than a handful of Silicon Valley monopolies. Blockchain is a general mechanism for running programs, storing data, and verifiably carrying out transactions. It is like a computer that's distributed and runs a billion times faster than the computer we have on our desktops, because it's the combination of everyone's computer.

The internet may be the foundation stones of the metaverse, but blockchain enables a network of security instead of vulnerability. The Web 3 metaverse promise is where everyone owns and controls their own data, safeguarding the enormous amount of data that power the idealized vision of the metaverse.[1]

If the metaverse is a mobile, living internet, then cryptocurrency is a mobile form of money. Blockchain is taken even further in the metaverse, though, and the function extends far beyond buying and selling, marketplaces are only one part of what the metaverse offers: China's metaverse boom, for example, relies heavily on online gaming. Social interactions, education, knowledge sharing. . . all these and more are functional by-products of Blockchain leveraged by the world builders of the metaverse. The decentralized, distributed network that is Blockchain is precisely the system needed to make the metaverse a user-centered, user-operated, and user-owned virtual community.

---

[1] The Metaverse as Virtual Heterotopia, David van der Merwe, 2021, www.socialsciencesconf.org.

Workflows

As with all emerging technologies, the exploration phase that metaverse, blockchain, and GPT are currently in is subject to success and failure measures that will cause debate and discussion. Our use of these technologies is being experimented with stand-alone solutions and as an ecosystem solution. We use a methodology of process mapping a workflow to identify what to automate (and what not to automate) and then insert the technology choice into the process. By using simple stand-alone solutions in this process, we can quickly identify success and failures. Once we can measure success as a stand-alone solution, we can then begin the process of creating an ecosystem that will also measure success and failures.

Our first attempt at integrating the emerging technologies of metaverse, blockchain and GPT as an ecosystem began with smart contracts in project delivery. U.S. Industry associations who currently control the marketshare of design and construction contracts are all exploring a migration to smart contracts. The American Institute of Architects (AIA Contract Documents) and the Associated General Contractors of America (AGC ConsensusDocs) are the market leading contracts used today. They both currently sell paper and PDF digital versions of their industry standard contracts. The use of blockchain smart contracts and their effects on delivery process is currently under exploration by both associations. We have provided a workflow example of how our organization could use this disruptive technology and its effect on the delivery process and subsequent effect on facility management.

Capital Asset Delivery Using Smart Contracts Workflow

Our first step creating a smart contract workflow was to develop a stand-alone solution. We created a contract between the architect and owner using ChatGPT. We found ChatGPT provided a good high-level series of suggested frameworks, but the technology fell short of providing detail-level information to properly develop a contract that would compete with the AIA or AGC. Once we consulted with legal advice, we settled on the wording and intent of our Owner/Architect contract. After testing as a stand-alone solution, we began the process of adding other workflow elements.

Construction Documents as the Digital DNA of the Built Environment

As the fulcrum of how all projects exist and are conducted, construction documents are the foundation elements of how the built environment conducts business. Usually, Construction Documents are how the designer communicates the design intent to the constructor on behalf of the building owner. As an industry, we have moved relatively fast from physical drawings and specifications as the construction documents to the emergence of digital drawings (CAD/BIM) and specifications as the new standard of care. As digital standards of care move from finite files to decentralized streamed legal documents through processes like Smart Contracts, our industry is yet again challenged to keep up with the times. The one element of stability as the workflow processes of designing, building, and delivering a building in today's world is that design professionals are more valued than ever as they produce the Digital DNA of the Built Environment (Figure 3-1).

Ethereum blockchain

After an analysis and testing of available blockchain solutions, we chose Ethereum blockchain to host our smart contracts. Our criteria for this decision involved:

- Security and Privacy.
- General Data Protection Regulation (GDPR) compliance.

Fig. 3-1: Expo2020 in Dubai Metaverse.
Courtesy: Paul Doherty.

- Consensus Mechanism.
- Network effect.
- Maturity of the solution.
- Operational costs.

Ethereum enables smart contracts built on its blockchain to run smoothly without fraud, downtime, control, or any third-party interference. Ethereum is also a programming language that helped us to create the smart contract functionality. On Ethereum, smart contracts are written in its Solidity programming language, which is Turing-complete. This means that the rules and limitations of smart contracts are built into the network's code and no bad actor can manipulate such rules.

Digital Twin

We took our construction documents (CDs – BIM and specifications) from a real-world project that was under construction and provided two tasks for our CDs. We put them in native format as a reference position to the smart contracts through an API (application programming interface). Simultaneously, we put our BIM into a gaming engine environment to enable a metaverse environment linked to our smart contracts. Our stand-alone solution was to have the native documents and the metaverse documents communicate with the smart contract. We found that the native documents using a homegrown API had superior value over the emerging metaverse solution. Ease of development, accuracy of the data/workflows, and integration into ecosystem solutions were the deciding factors. As we define a digital twin as a mirror image between the physical and digital environments, leveraging this definition into a smart contracts process made the native file format an easy decision (Figure 3-2).

Geo Location and Workflow

We then required that all building materials furniture, fixtures, equipment, and appliances that enter our project site had to have a form of geo-location. GPS, RFID, Bar Code, QR Code, etc. are all accepted. Using API technology, we can acquire geo-location data as its being put in place in real-time and have this data communicate with our BIM. Our BIM uses this geo-location information to determine if the asset was put in place according to the construction documents. We added the element of the project schedule to provide the data if the construction task was completed on time. We then added an API

Fig. 3-2: Qingdao, China Virtual Reality Theme Park Metaverse.
Courtesy: Paul Doherty.

to construction performance to accept the on-site report from the project manager if the task was performed at the proper quality level. If all of these criteria are met, a decision application was implemented to release payment online to pay the contractor/sub-contractor, creating trust and loyalty relationships. For the Owner, the Lien Release process happens quicker than traditional processes.

Facility Management

A discovered output was that because we were collecting real-time data of the placement of building elements based on our smart contract, we were creating an as-built/record document with amazing accuracy as the asset was being created. This resulted in an as-built/record document that became an instant asset for facility management. Integrating FM documents into the world of blockchain and the metaverse we are now exploring what the new opportunities are exposing themselves.

Challenges

The promise of the emerging technologies of metaverse, blockchain, and GPT has strong challenges, some of which we are facing directly today. Some of our failures were due to technological limitations, while others were due to human limitations. We want this book to provide guidance on what we have found to be valuable in our work and what we feel will be valuable for the market.

We identified a small pilot project to implement our metaverse, blockchain, and GPT technologies. It is a 50,000 square foot commercial office building that we built, own, and operate. We had full internal control and ownership of decisions for workflow and technology selection. We used our own developed metaverse using the Unity gaming engine; we used the Ethereum blockchain; and we used ChatGPT as our GPT solution. We used the initial data capture from our Construction Documents (BIM Revit, Specifications Microsoft Word) and our data capture from our operational facility management solutions that included TRIRIGA, Maximo, Archibus, and Siemens solutions. We had a pilot project team that included 4 full-time employees, 2 consultants, and 8 part-time subject matter experts who are involved in our process for scheduled tasks. The pilot project began in April 2022 and is scheduled to end in April 2024. There has been an expanded program to move beyond the pilot project in May 2024 that has an indefinite end but is measured by return on investment and other business planning measures (Figure 3-3).

Our initial measures for the pilot and looking forward to the expanded program, we have found initial challenges that our metaverse, blockchain, and GPT solutions have to answer. Our pilot project provided key learnings as to the acceptance and adoption of metaverse, blockchain, and GPT as an ecosystem solution for a single commercial building.

Fig. 3-3: Qingdao, China Virtual Reality Theme Park Metaverse.
Courtesy: Paul Doherty.

Governance in a Decentralized Digital Environment

Our ecosystem technology solution works best as a data-centric driven workflow in a decentralized digital environment. Using distributed ledger technology (DLT) technology in the form of blockchain, our single building solution was met with initial skepticism from leadership and management. A decentralized digital environment is a type of digital ecosystem that is not controlled by a single entity or organization. Instead, it relies on a network of decentralized nodes, or participants, to store, transmit, and process data. There are attempts to describe this governance as a Distributed Autonomous Organization (DAO). This type of environment is often associated with blockchain technology, which uses a distributed ledger to record transactions and data in a way that is secure, transparent, and immutable. Decentralized digital environments can offer a number of benefits, including increased security, privacy, and autonomy, as well as the ability to facilitate peer-to-peer interactions and exchanges without the need for intermediaries. Our challenge was to convince all stakeholders in the pilot that a DAO was the proper governance process, when we, as the owner, really have full control of the pilot. We have a skeptical work environment as to the acceptance of the DAO governance, including from our own internal team. Changing past processes and comfort zones from individuals and organizations is our greatest challenge for this pilot.

Cybersecurity

Our Pilot Project was met with cybersecurity challenges from the inception of the project. The following issues were addressed:

- Phishing attacks: These are fraudulent emails that are designed to trick people into revealing sensitive information, such as passwords or financial information. We have had no phishing attacks on our network during the pilot.
- Malware: This refers to software that is designed to cause harm to a computer system, such as viruses, worms, and ransomware. We have had no instances of malware during the pilot.
- Unsecured networks: When a network is not properly secured, it can be accessed by unauthorized parties, who can then access sensitive data or disrupt our network. We found holes in our network during our pilot testing. The holes have been secured.

- Weak passwords: Using weak passwords, or reusing passwords across multiple accounts, can make it easy for attackers to gain access to sensitive information. We relied on education to our team on adhering to this threat.
- Insider threats: Insiders, such as employees or contractors, can pose a threat to an organization's cybersecurity if they have access to sensitive data and intentionally or unintentionally misuse it. We implemented a sabotage identification system that provided no positive results during our pilot.
- IoT vulnerabilities: The increasing number of internet-connected devices, such as smart devices, can create new vulnerabilities if they are not properly secured. We made our IoT equipment suppliers provide secure elements before installation.
- Lack of employee education: If employees are not properly trained regarding cybersecurity best practices, they may unknowingly expose us to risk. We provided both physical classroom-style education, online education, mobile app education and delivered periodic audits to assure compliance.

Trust

Trust is an important aspect of relationships, both personal and professional. It refers to the belief in the reliability, truth, ability, or strength of someone or something. When we trust someone or something, we feel confident in their ability to do what they say they will do, or to be truthful and honest. Trust is built over time through consistent, reliable behavior and communication. It is an essential foundation for healthy relationships, as it allows us to feel safe and secure in our interactions with others. Trust can be broken if someone acts in a way that goes against what we believe about them or if they betray our trust in some way. Maintaining trust in a relationship requires effort and commitment from all parties involved. Our biggest challenge was building trust between people in the project. We learned that time together between people can be a positive element in building trust between people. This process can also validate distrust between people, making company leadership and management decisions an easier and more valuable exercise. To address trust issues in the AEC/FM industry, it is important to establish clear communication channels, promote transparency, define roles and responsibilities, and work to build strong relationships between stakeholders.

Data

Using smart contracts and GPT solutions inside a single building pilot, we provided our design/construction team and facility management team trusted data. The issue of trust was difficult to achieve in our single building pilot. The traditional issues of trust between stakeholders became apparent when we asked the different stakeholder parties to trust the information provided by process parties. We needed to insert workshops to educate the stakeholders as to the authenticity and trustworthiness of the data provided during the design and construction process that would result in accurate as built/record documents to be used for facility management purposes. We felt that we satisfied most project stakeholders, but that many were not convinced that the data we provided was authentically accepted.

Avatars

The introduction of avatars as stakeholders and role players in an active project proved to be very difficult. Avatars that were representative of who they were in real life proved to be more effective than avatars who were not representative of who they were in real life. We found that a team member's experience with gaming and gaming environments was a major contributor to the adoption of team members use in the project as an avatar. As an organization, we realized that we need to provide a detailed value proposition of the use of avatars to the project stakeholders and role players to gain better understanding and adoption.

Smart Contracts

Gaining the trust of all stakeholders to adopting the use of an online smart contract had a strong resistance to change. Some team members and stakeholders were resistant to using a model-led workflow, especially if they are not familiar with the technology or processes involved. It took time to overcome this resistance and to get everyone on board with the new workflow. Our primary contracts were smart contracts, while some subcontracts had to remain in the traditional physical document process.

Value Propositions

Our emerging technology research of metaverse, blockchain, and GPT will provide value over time. I wanted to explore the ecosystem approach on my projects and discovered that the technology is still emerging and immature, resulting in limited value. But as we continue our journey with continuous innovation improvement, emerging technologies become mature market technologies. Facility management has a very bright future with these innovations.

Technology has had a significant impact on facility management, and it continues to shape the way facilities are managed and maintained. Some of the value of metaverse, blockchain, and GPT technologies on facility management that we found include:

Increased Efficiency

Metaverse, blockchain, and GPT technologies assisted our facility managers to streamline processes and improve efficiency. For example, using building management systems and energy management software in a smart contract and metaverse environment helped monitor and control building systems remotely, reduce energy consumption in one building by 3% year-on-year, and optimize our maintenance schedules.

Improved Data Collection and Analysis

GPT technology is beginning to assist our facility managers in collecting and analyzing data to make informed decisions about the maintenance and operation of our facilities. For example, using sensors and other data-gathering tools, we feed this data into our GPT system that helps our managers track energy consumption, identify maintenance issues, and optimize operations. We have learned that we need to expand our data collection to more than one building to realize the full potential of GPT as our results were limited. By expanding our GPT use to our portfolio of buildings, we expect more robust analysis and reporting to improve tasks like enhancing our customers' experience by having more intelligent mobile apps and virtual assistants (Figure 3-4).

Accurate and Trusted Facility Data and Information

Our Smart Contracts and metaverse solutions provided transparency that resulted in an accurate and mutually trusted environment for our single building pilot. As stakeholders saw the results of the use of Smart Contracts, the initial skepticism of the trust of the data began to decrease and the trust began to increase. This energy had some

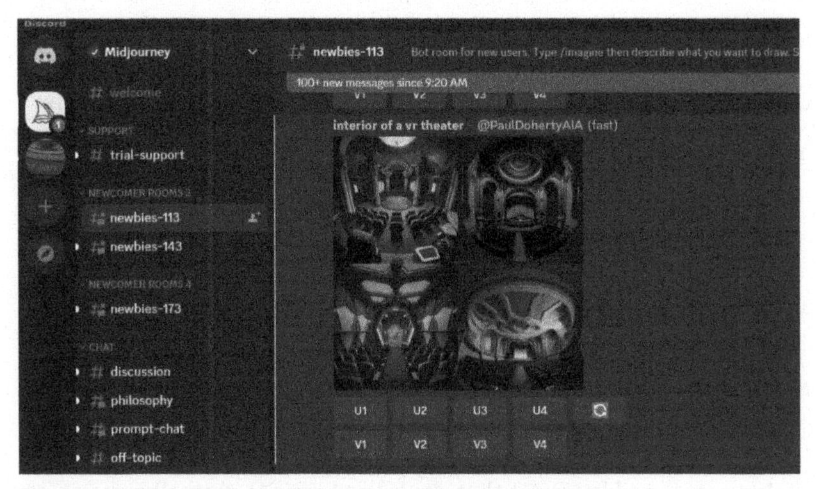

Fig. 3-4: GPT-style generated graphics using Midjourney on Discord.
Courtesy: Paul Doherty.

stakeholders providing their own ideas of how our team could use the trusted data for better communications, more accurate analysis, and more informed decisions.

Tokenomics

Tokens are programmable money. The term actually comes as a metaphor to gaming arcades, where you need a token to play the game. Tokenomics may potentially be used by facility management as a means of incentivizing or rewarding certain behaviors or actions related to the maintenance and operation of a facility. For example, working with our HR team we issued tokens as a reward to individuals who contribute to the maintenance or improvement of a facility. These tokens were then used to redeem goods or services, while some people traded these tokens on a cryptocurrency exchange. This began building a system of incentives that encourages people to actively contribute to the management and maintenance of the facility. We plan to expand this system to more of our buildings as we complete the pilot test period.

Our continued research will refer to our learned knowledge of our initial single building pilot as valuable benchmarking. Our implementation and deployment of metaverse, blockchain, and GPT technology will be dictated by internal project adoption and industry market acceptance. Since most of our research and development for our initial pilot project was very innovative and did not have many referenceable sources for this book, it is our hope that we can find others that are innovating in the same direction in order to provide benchmarking and information sharing opportunities for the greater community. Our projects in South Korea, China, Saudi Arabia, and the USA will continue to be sources of analysis and reporting of the adoption and implementation of metaverse, blockchain, and GPT technologies that we will continue to share with the world.

Woven Collisions: NFTs and the Metaverse

Another use of DLT/Blockchain technology is the process of creating and exchanging Non-Fungible Tokens (NFTs). NFTs are unique digital files whose authenticity is verified by the blockchain, an ever-expanding digital ledger that records all sorts of digital transactions. Being non-fungible means that the tokens are not mutually interchangeable. Unlike Bitcoin, you cannot exchange one NFT for another as each is unique. They cannot be divided into smaller portions like most currencies or precious metals.

The cryptographic origins of NFTs ensure their digital scarcity and proof of ownership. NFTs can be used across a variety of applications, such as digital art or collectibles. This type of item is of particular interest to digital artists and IP Owners as it prevents the art or collectible from being endlessly copied and a specific piece of digital art or collectible can have a verifiable chain of custody. In some cases, the artist (IP Owner) can collect royalties on an item that passes between owner's multiple times (Figure 3-5).

Behind the cumbersome term lies the technology based on the Ethereum blockchain that transforms digital objects into unique and 100% counterfeit-proof collector's items.

One of the beautiful things about public blockchains like Ethereum is the ability to digitally own an immutable, transparent, and distributed asset. The advent of token standards like ERC721 allows us to customize these ideas even further to create scarce, liquid tradable goods on a global scale.

At TDG, we are leveraging NFTs for the built environment in the form of:

- The design aesthetic of a building
- The elements that make up the building (NFTs for materials, systems, equipment, appliances) that can be measured for both work in place and lifecycle performance

Fig. 3-5: NFT galleries in the Metaverse.
Courtesy: Paul Doherty.

- Creating both UN Sustainable Development Goals (SGD) and Environment/Social/Governance (ESG) measures for active NFTs
- NFTs as an element for all Smart Contracts in the Construction Industry

The immediate use of NFTs that we have implemented is to develop virtual galleries for the display and transactions of NFTs by celebrities, iconic collectibles, and luxury merchandise. This woven collision of NFTs being experienced in the metaverse is just the beginning of what we see as a decentralized metaverse that has infinite experiences all experiencing a world of digital real estate.

Another area of interest regarding the use of the metaverse in the design and construction process is the use of 24 × 7 artificial intelligence (AI)-driven avatars as an extension of decision-making elements for successful project delivery. In the modern gaming world, the game never ends and is always on. With Minecraft, Roblox, and Fortnite as examples, players can use their avatars for playing their games in real time and once they leave the game, the game is still being played/experienced by others. The opportunity for construction projects in adopting avatars as "players," is that, like a game, the construction project never stops. There could be fewer workers at certain times of the day, but the project is ongoing. By adopting a metaverse environment as a component of project controls, avatars of each stakeholder can continue to work in some capacity like analysis, reporting, decision making, etc. even when the physical stakeholder is no longer on site. This AI "helper" can also be the conscience of the project, enabling and suggesting guidance to the project team to improve quality and safety, decreasing time and saving money. This use of the Industrialized Metaverse means that issues can be found before they arise, analyzed, and fixed quickly. What we find so exciting is that our use of the Industrialized Metaverse is where we can travel into the past and the future to understand issues and processes better, while finding the proper solutions.

LOE

There is an immediate need to categorize the Metaverse into Levels of Experience or Levels of Engagement (LOE) in order to design these digital environments properly and have users not get frustrated in navigating the Metaverse. What works well in the physical world with a building design can be a disaster in the Metaverse. So simply taking a BIM and "saving as" into a gaming engine is not best practice, and in certain cases, can lead to unwanted legal issues. Who owns the rights to the Digital Twin (the designer, the building owner, etc.) is a lawsuit waiting to happen. So do not fall for the Silicon Snake Oil salesman who promises to unlock your design firm's untapped value of taking your BIM to the Metaverse. Best practice dictates that taking proper steps with ownership rights management, defining the LOE to be used in each instance, and learning overtime how the digital asset should be designed with rich experiences that cannot be designed in the physical world (like the ability to turn on and off physics). Physical architecture may have a lot to learn from its digital twin and subsequent LOE in the Metaverse.

Real-Estate-Backed Digital Asset Securities

At TDG, we are using DLT to create fungible tokens that act as financial security for our digital assets, like BIM. The intrinsic value of the physical real estate asset acts as our digital twin financially, providing a robust trading market for our digital assets. Real-estate-backed digital securities.

We are developing a classification of securities that will move beyond BIM being the only digital asset. DLT will allow us to "nest" digital assets of various classes of securities for a single physical asset, thus allowing the intrinsic value of the physical asset to be used as the foundation of the digital asset. Any transactions (licensing, etc.) of the digital asset then builds the value of that digital asset. In time, the real-world value of a digital asset becomes more valuable than the physical real estate it represents. Other measures we are currently developing for digital assets on the blockchain will lead to innovative financial vehicles that we are benchmarking against current securities markets like futures, digital credit swaps, and ETFs that can be backed by Central Bank Digital Currencies (CBDC) or crypto currencies.

We do this by using our building systems as NFTs. Mechanical, Electrical, Plumbing and Information Communication Technology (ICT) systems being the 4 primary NFTs for each of our buildings. By using a smart contract to measure each NFTs work measure, we now have a real-time system in place that can provide queries as to certain measures we wish to have reporting of this trusted data. We then place this data and the 3D geometry model onto a trusted digital globe, like Cesium (www.cesium.com), that places the 3D digital model and its data assets onto a digital twin style in a latitude/longitudinal location. We then tokenize the 3D model with NFT data as a Fungible Token, securitizing the digital asset. This complete digital asset with a geolocation on digital planet earth can now be sold as a fractionalized asset in a trusted digital globe with trusted operational data. The fractionalized ownership of each digital asset is then value-based on the IRL building's performance. Profit/Loss performance data being a primary measure of how each physical asset is being valued in the digital asset environment. True digital real estate, not the Decentraland/The Sandbox models. The distributed ledger portion for IP & then pinning it to the value of an underlying real estate security is perfectly viable and has an exciting future.

Web 3

Web 3, also known as the decentralized web, is the next evolution of the internet, characterized by the use of blockchain technology and decentralized protocols. It is a vision of a new internet that is more secure, transparent, and user-centric, where users have greater control over their data and can interact with each other and digital services without relying on centralized intermediaries.

In the Web 3 ecosystem, applications and services are built on decentralized protocols and run on peer-to-peer networks, allowing for greater privacy, security, and censorship resistance. Blockchain technology is a key component of Web 3, providing a trustless, tamper-proof record of transactions and interactions. Web 3 also enables the creation of decentralized autonomous organizations (DAOs), which are self-governing entities that operate according to predefined rules encoded on the blockchain. These organizations can provide new models for collaboration and governance and will disrupt traditional centralized business models, including the businesses of the built environment.

Web 3's effect on design and construction companies includes improved supply chain management, which is one of the most promising use cases for blockchain technology in construction. By using blockchain to track the movement of materials and goods throughout the construction process, companies can improve transparency, reduce waste, and prevent fraud. Web 3 will also improve the security of project management tools used in construction. Decentralized platforms could enable more secure collaboration between contractors, architects, and other stakeholders involved in the construction process. Web 3.0 will enable greater efficiency and automation in construction. Smart contracts, for example, are being tested to automatically trigger payments when certain project milestones are reached, reducing the need for manual tracking and payment processing. And perhaps the largest value is with enhanced building performance analysis and reporting. With the use of IoT devices and blockchain-based data management, construction firms could better monitor and manage building performance throughout the lifecycle of a project, improving the overall sustainability and resiliency of buildings.

Once the building is built, Web 3 brings exciting solutions to facility managers in the form of smart contracts. One of the primary benefits of Web 3 is the use of smart contracts, which are self-executing contracts with the terms of the agreement between buyer and seller being directly written into lines of code. This technology can be used in facility management to automate various processes such as lease agreements, work orders, and maintenance contracts, resulting in increased efficiency and reduced costs. Asset tracking is another big winner with Web 3 solutions. Blockchain technology can be used to track the entire lifecycle of assets, including equipment and inventory, making it easier for facility managers to keep track of their assets, ensure they are being used efficiently, and prevent theft or loss. One of my favorite Web 3 solutions for FM is decentralized energy management. Web 3 is being enabled for decentralized energy management, where energy generated from renewable sources, such as solar panels or wind turbines, can be tracked and traded through a decentralized system. This could result in reduced energy costs for facility managers, as well as increased transparency and accountability. An area of constant concern is data security. Web 3's decentralized architecture makes it more secure than the traditional centralized internet. By storing data across a network of nodes rather than in a single centralized location, facility managers can benefit from increased security, data integrity, and reduced risk of data breaches.

Web 3 has the potential to revolutionize design, construction, and facility management by providing new tools and solutions for managing assets, automating processes, and enhancing collaboration, all while improving data security and reducing costs.

AI

Artificial intelligence (AI) and related technologies are beginning to play a major role in the design, construction, and facility management process.

One way in which AI can affect building design is through the use of computer-aided design (CAD) and Building Information Modeling (BIM) software. These tools can help architects and designers to create and visualize buildings in a more efficient and accurate way. They can also help to optimize building performance, such as energy efficiency and structural stability, by simulating various scenarios and identifying potential problems before construction begins. AI can be used to analyze large amounts of data to help inform building design decisions. For example, machine learning algorithms can be used to analyze data on building usage patterns and environmental conditions to optimize energy usage and occupant comfort.

In terms of construction, AI and robotics can play a role in automating certain tasks, such as site preparation, material handling, and quality control. This can help to improve efficiency, reduce costs, and improve safety on construction sites. AI and related technologies can play a significant role in improving the efficiency, accuracy, and sustainability of the building design, construction, and facility management process.

AI has the potential to greatly impact facility management by improving operational efficiency, optimizing energy consumption, enhancing security and safety, and enabling predictive maintenance. In the case of predictive maintenance, AI can analyze data from sensors and machines to predict when maintenance is needed before equipment fails, reducing downtime and maintenance costs. AI can also analyze energy consumption patterns and suggest ways to optimize energy usage in facilities, leading to cost savings and reducing carbon emissions. A large value of AI in facilities is with safety and security. AI-powered video analytics and facial recognition can enhance security and safety in facilities by identifying potential threats and improving access control. Concerning space utilization, AI can analyze occupancy patterns and suggest ways to optimize space utilization, leading to cost savings and improved productivity. And regarding smart buildings, AI can enable the integration of various building systems such as lighting, HVAC, and security to create smart buildings that can be controlled and optimized in real-time (Figure 3-6).

AI also has the potential to make the real estate industry more efficient, transparent, and accessible to buyers and sellers alike. Tasks like property valuation are being directly affected. AI can be used to determine the value of a property based on a range of factors such as location, size, amenities, and nearby infrastructure. This can help real estate agents and buyers make more informed decisions about pricing. Another task affected is property search. AI-powered property search engines can help buyers find properties that match their specific criteria. These engines can use natural language processing to understand the buyer's preferences and return the most relevant listings. AI can also assist property managers in tasks such as maintenance scheduling and tenant screening. This can help reduce costs and improve efficiency.

Fig. 3-6: I asked AI to mashup Batman and myself and this was the result. I don't like AI at the moment.
Courtesy: Paul Doherty.

GPT

GPT (generative pre-trained transformer) is an artificial intelligence language model that is primarily used for natural language processing tasks such as text generation, translation, and summarization. GPT is having an impact on construction with tasks like project documentation. GPT can be used to generate project documentation such as reports, contracts, and proposals. This can save time and reduce errors associated with manual documentation. Another GPT value is in communication. Construction projects involve a lot of communication between various stakeholders such as architects, contractors, and clients. GPT can be used to analyze and summarize communication data, which can help improve communication and collaboration. An area years in the making using GPT is Planning. GPT can be used to analyze data from past construction projects to identify patterns and trends. This can help construction companies to plan projects more effectively, reduce costs, and improve efficiency. And most importantly, GPT and Safety. GPT can be used to analyze safety data from construction sites and identify potential hazards. This can help to prevent accidents and improve safety on construction sites.

GPT has the potential to improve various aspects of the construction industry and make it more efficient and effective. GPT also has the potential to revolutionize the real estate industry by automating many repetitive tasks and providing valuable insights for real estate professionals. One example is natural language processing for real estate data analysis. Real estate is a data-driven industry, and GPT can help in natural language processing for analyzing data, identifying trends, and forecasting market changes. Real estate customer service could use chatbot technology that can provide 24/7 customer service to potential buyers and sellers by answering their queries in real-time, which can increase customer satisfaction and save time for real estate agents. Another example is automated property descriptions. GPT can generate automated property descriptions based on real estate data, which can save time for agents and make it easier to create consistent and appealing property listings. GPT can also analyze large datasets to predict property values based on various factors such as location, market trends, and property features, which can help real estate professionals to make better decisions.

GPT is a type of language model developed by OpenAI to be the product ChatGPT that uses a transformer architecture to generate human-like text. It is trained using a large dataset of human-generated text, such as books, articles, and websites.

To train GPT, the model is fed a large dataset of text and is then asked to predict the next word in a sequence based on the words that come before it. The model is then evaluated on its ability to accurately predict the next word, and its parameters are adjusted accordingly to improve its performance. This process is repeated many times until the model reaches a satisfactory level of performance.

GPT is trained using unsupervised learning, which means that it is not given explicit correct answers or labels for the data it is trained on. Instead, it learns to generate text by identifying patterns and structures in the data on its own. This allows GPT to generate a wide range of text, from simple lists to more complex narratives and articles.

ChatGPT Model

ChatGPT is a variant of the GPT language model that is specifically designed for generating text in a conversational style. It was developed by OpenAI and was trained on a large dataset of human conversations in order to learn how to generate text that sounds natural and engaging in a conversational context.

The ChatGPT model works by taking in a given input, such as a question or statement from a user, and using the pre-trained transformer network to generate a response based on the input. The model is able to understand the context and meaning of the input and generate a response that is coherent and appropriate in the chatbot conversation context.

One of the key features of ChatGPT is its ability to generate long and complex responses, which allows it to hold more natural and engaging conversations with users. It is also able to handle a wide range of topics and styles of conversation, making it a flexible and powerful tool for chatbot developers.

Open-AI researchers and engineers trained ChatGPT with a machine learning technique known as deep learning. Deep learning is a type of machine learning that attempts to mimic the human brain by recognizing patterns in a large amount of data.

An example of the value of GPT to our industry is its use in writing construction specifications. By using a template like the Construction Specification Institute's Master Format, taxonomies and sub-taxonomies can train GPT to vertically integrate and learn best practices from a data lake, ultimately delivering an accurate written specification in seconds. Why this is important, especially for industry professionals like architects and engineers, is that their business model of billable hours just became obsolete due to the speed of writing construction specifications using GPT. A different and more relevant business model must be created based on value, not billable hours.

2023 witnessed the emergence of AI and GPT innovations at an amazing velocity. GPT has already led to human jobs being lost to this technology in the fields of customer service, social media management, financial advisement, and paralegals. In our industry, we expect back office, labor-intensive functions like quantity take offs, specification writing, RFP responses, change order management, among many others, to be affected by GPT and AI in a short amount of time. This will give rise to a cottage industry of online specialty AI/GPT services that will replace the amount of workers necessary to design, build, and manage our built environment.

AI/GPT will not take your job, but someone using AI/GPT absolutely will.

Now that you have a snapshot of the metaverse and its mechanisms, let's explore a view of the future using a Crystal Ball and guidance from some great minds in our industry.

CHAPTER

4

The Crystal Ball

As the world of the metaverse emerges, there is no clear path as to when and how elements will come together to create the Utopian vision of the metaverse. I personally hope that a Utopian version of the metaverse does not emerge, but rather a more pragmatic version emerges over a period of time as an evolutionary developed environment. I do not see a Big Bang moment that states that the metaverse has arrived. I do see elements of the metaverse having Big Bang moments, like the National Basketball Association (NBA) showcasing during the 2023 All Star Game, an avatar-morphing AR technology that allows an image of you to morph into an NBA star's moves of making a play. That opened a lot of eyes around the world as to the possibilities of AR and Avatar technology. I expect to witness many metaverse elements have their time in the sun over the next few years. With each element maturing, it feeds into the metaverse ecosystem that lifts the quality of the user experience that becomes woven into a user's daily life.

I can vision a metaverse future that has cyber–physical interactions between humans, their physical environment, and a digital medium that has layers of experiences that humans can engage with. The industrial metaverse layer will focus on a person's work life, while a social metaverse layer will allow for live, play, and learn environments, while a personal metaverse layer will provide private environments for relationships, with oneself, with family, with friends, and for personal explorations. These explorations can be for fun, education, or curiosity. Since the metaverse requires its inhabitants to have a digital avatar, which can be a digital twin of your existing self or you can be someone or something else, imagine the opportunity to become an avatar of someone you want to explore of being them. Imagine you are an American white male, college educated, upper middle class, living in an American city IRL. You have the curiosity that you want to learn about being a minority in an American city and enter a virtual world as a minority avatar. Can this person learn what it is really like to be a minority in this metaverse setting? Of course not. But can some lessons be learned that can provide thought to this person as they live their life? Potentially yes. Avatars are vehicles in the metaverse whose outcomes are yet to be realized, but there is the hope that they can become vehicles of discovery, friendship, change, and common purpose (Figure 4-1).

Fig. 4-1: AI-generated graphics in the Metaverse.
Courtesy: Paul Doherty.

The idea of common purpose is a powerful force that we as a company have embraced to give a vision and a mission to our smart city mega projects around the world. Since we create a digital twin and eventually a virtual world in the metaverse with every project, this common purpose force is not just integrated into our physical assets, but also our digital assets. It is my firm belief that due to the environment of the metaverse, the world's next great religion will be born in the metaverse. To me, this is both interesting and terrifying at the same time.

What gives me hope beyond the religious element is that a spiritual environment emerges as the ethos of the metaverse, which brings up our company's ethos that in our projects there is the dichotomy of a scarcity vs. an abundance mindset. And this physical situation becomes a digital manifestation in our use of the metaverse.

Scarcity and Abundance

Abundance is a global phenomenon that requires us to rethink our way of producing, consuming, and living. An abundant life consists of an abundance of love, joy, peace, and just not an abundance of "stuff." When you feel grateful and appreciative, everything in your life feels like more than enough because you are appreciative of all the things you do have and you don't feel the desire for more. When you feel like you have enough or even an overflow, that feeling is abundance.

The main differences between an abundance mindset and a scarcity mindset are simple. With a scarcity mindset, everything is getting smaller. But with an abundance mindset, everything is getting exponentially bigger. This fits perfectly with the digital environment of the metaverse which can be designed to be exponentially bigger in all regards. Abundance mindset also promotes a win–win mentality, if there's enough for everyone, there's no need to compare and compete. You don't have to push someone down to raise yourself up. You can get what you want without snatching a piece of the pie from others.

Scarcity mindset leads to the belief that there are limited opportunities, options, and resources, while abundance mindset tells you that there are enough resources and successes for all to share. Thinking this way makes winning in corporate environments equivalent to beating someone else. In order for you to succeed, someone else must fail. In order for you to get what you want, someone else must have less of what they want. On the other hand, an abundance mindset flows out of a deep inner sense of personal worth and security.

Fig. 4-2: An Abundance of Plenty.
Courtesy: Paul Doherty.

It's grounded in the belief that there is more than enough for everyone. While in the metaverse, there are exercises you can open your mind to include:

- Notice and reframe scarcity-based thoughts
- Train your mind to recognize the possibilities, not the limits
- Practice mindfulness
- Practice gratitude (Figure 4-2).

Not everyone thinks this way though. Those who view the world as abundant find solace in the knowledge that there are enough opportunities for everyone out there. Instead of hurrying through their decisions, they make deliberate and careful choices aligned with their values and the life they want to live. By bringing this mindset to the metaverse, the metaverse environment has the opportunity to shift mindsets both in the digital world and IRL.

Edge Computing

The metaverse is a virtual universe where people can interact with each other and digital objects in real-time. Edge computing is a distributed computing model that brings computing and data storage closer to the end-user, which can significantly reduce latency and improve the performance of applications.

Edge computing can have a significant impact on the metaverse. By processing data closer to the end-user, edge computing can reduce the latency associated with transmitting data between the user and the metaverse's servers. This reduced latency can improve the user's experience by reducing lag and increasing the responsiveness of the metaverse's applications.

Additionally, edge computing can enable the development of new applications and experiences within the metaverse. For example, edge computing can enable the creation of immersive and interactive experiences that rely on real-time data processing, such as virtual reality (VR) and augmented reality (AR) applications. These applications require real-time processing of large amounts of data, which can be challenging to achieve with traditional cloud computing architectures.

Edge computing can significantly enhance the performance and capabilities of the metaverse by reducing latency and enabling the development of new applications and experiences. As the metaverse continues to evolve and grow, edge computing is likely to play an increasingly important role in shaping the development of the physical and digital worlds.

The superior performance on reducing latency in virtual worlds has made edge computing a foundation in the metaverse's creation with many industry insiders. Multi-access edge computing (MEC) is expected to boost metaverse user experience by providing standard and universal edge offloading services one click from mobile-connected user devices.

This means the design and construction industry will have a major role to play in the advancement of the growth of the Metaverse as the adoption of integration of edge computing into our built environment will become a necessity, leading to the recognition of Information Communications Technologies (ICT) as a Fourth Utility of all buildings, joining Mechanical, Electrical, and Plumbing utilities.

To optimize the interaction between the cloud and the edge, an efficient orchestrator is a necessity to meet diversified and stringent requirements for different processes in the metaverse.

Is this the new role for architects? Buildings as computers in the edge realm, holding vast amounts of trusted data like a server bank, that allow the changes, the delta's to be the data that is transacted (Figure 4-3).

Fig. 4-3: A TDG Metaverse.
Courtesy: Paul Doherty.

Censorship

Once the metaverse becomes a popular place for content creations, numerous user interaction traces and new content will be created. For instance, Minecraft has been regarded as a remarkable virtual world in which avatars have a high degree of freedom to create new user-generated content. Minecraft also supports highly diversified users who intend to meet and disseminate information in such virtual worlds. In 2020, Minecraft acted as a platform to hold the first library for censored information, named The Uncensored Library, with the emphasis of "A safe haven for press freedom, but the content you find in these virtual rooms is illegal." Analogous to the censorship employed on the Internet, we conjecture that similar censorship approaches will be exerted in the metaverse, especially when the virtual worlds in the metaverse grow exponentially, for instance, blocking the access of certain virtual objects and virtual environments in the metaverse. It is projected that censorship may potentially hurt the interoperability between virtual worlds, e.g. will the users' logs and their interaction traces be eradicated in one censored virtual environment? As such, do we have any way of preserving the ruined records? Alternatively, can we have any instruments temporarily served as a haven for sensitive and restricted information? Also, other new scenarios will appear in virtual 3D spaces. For example, censorship can be applied to restrict certain avatar behaviors, e.g. removal of some keywords in their avatars' speeches, forbidding avatars' body gestures, and other non-verbal communication means.[1]

Thought Leader Interviews

To contribute to a Crystal Ball chapter of this book, I thought it would be good to have other respected voices provide their takes and viewpoints that can provide both validation and challenges of many of the concepts in this book. The tone of this part of the chapter will be in an interview/interviewer format. I am providing experts in the Metaverse, Data, Process, and the Future. I hope you enjoy this part of the chapter as much as I did the interviews and learn from these great minds.

Damon Hernandez – The Metaverse

Damon has been my friend and my trusted advisor for all things metaverse for over a decade. Since I've known Damon, one of his social media accounts is named "metaverse one" as an example of how far ahead of the crowd Damon has been and continues to be. I have a deep respect for his opinions and I am thrilled to share this interview with you.

PAUL: Thanks for agreeing to share your insights and views regarding the metaverse, Damon. I want to section our conversation into four big areas that will mimic this book: (i) Digital twins, virtual worlds and the metaverse, (ii) Blockchain, smart contracts, and digital real estate, (iii) Web 3, GPT, and AI, and (iv) a Blue Sky/Crystal Ball of your vision of the future.

DAMON: Thanks for having me. I think the most important thing that stands out is that we must have thought of the metaverse in a holistic way. The metaverse unfortunately is not going to become a reality in the way many individuals think it will.

---

[1] All One Needs to Know about Metaverse: A Complete Survey on Technological Singularity, Virtual Ecosystem, and Research Agenda. Journal of Latex Class Files, Vol. 14, No. 8, September 2021.

I still feel we're easily a good 10 years out. And even with that 10-year horizon line, we are building the idea of the metaverse not for you and me, but for the upcoming generations. As an example, I started doing these TikTok postings talking about technology recently. My 16-year-old nephew talked with me about this. He is living in these virtual worlds where they pop in and out as a complement to the physical world. So, it's not even a digital reality. It's just a part of their reality. It's like, Hey, I'm going to go hang out and it doesn't matter if it's in a virtual world or physical world.

So, 100%, I'd say the metaverse is not going to be where I'm going to be going to work unless I'm working longer than I want to be. We're building it for the next wave in that my nephew is going to be one of my guests. Actually, he and I are going to be like co-hosts because it is his generation. Because of this, I want to hear from the generation that's coming up. What are their thoughts about privacy in the metaverse, have it from their perspective. Our time spent today is that we are building the metaverse for your son, not for us, would be the first thing I would say.

Another point regarding the adoption of the metaverse is that every country is different as far as that adoption goes. I think I think the United States is still beautifully stuck in the 20th century and a lot of its thinking and I attribute a lot of that to just the people that are in charge and their age. The technology of talking about it in Europe is a completely different conversation than the US. I mean, yes, the markets are smaller, so it's a lot easier for technology to be adopted faster. But in some cases, I think it's just about adoption and technology and saying what is the value of this type of experience?

Part of the problem is that experts in one innovation do not look beyond their expertise to see how it all connects with others. Where are the environments for true collaboration with these new innovations? Many people freak out when some of these innovations like AI, VR, AR, and others make a market impact. As a developer, I raise an eyebrow to say, these innovations are already in my spice cabinet as ingredients for my solutions. What's the problem? No one should care about the spices I use; it is about how they come together and what they make right. So many people do not think or understand an ecosystem approach to technology, which brings us to today's dilemma of having the clueless leading the new technology or the new technology leading the clueless, I'm not quite sure which one it is.

All of this leads to real-world questions of why do I care about the metaverse? How's the metaverse going to add value to my life? And then I think just like the web, there's not going to be one owner. Now you'll have things that seem like they're an owner of the metaverse or something like that, just because like now most people would say, well, the web is Facebook, the web is Google. I mean, it is and it isn't. So, I think that in that similarly are these things with this idea of interconnected virtual worlds is that you're gonna have some virtual worlds or experiences that have super large populations. But then it's about that interconnectivity. So, I don't know if we'll get there anytime soon, in that sense of interconnected virtual worlds, because of the business models like that was the problem with 2006 to 2009, virtual worlds and all these people were talking about Metaverse, Metaverse, Metaverse, and it's not technology. It's more of show me a business model that allows you to add value in one place, and then have that person take maybe that value out of your platform into another platform. Then even if you solve those issues of business model interoperability, then I still think it's the whole, as I've mentioned before, then what you'll run into is what are the rules of this space as governed by the participants of the community? And it should be my right as a person who owns Star Wars Stormtrooper armor if I did, to be able to wear what I want, where I want, and to be able to take that with me, because that's my asset that I want to take to another virtual world. But if I show up at a renaissance fair as a Stormtrooper, I've pissed off a lot of people. The other issue of this interoperability of how do we go smoothly from one world to the next? And, and unfortunately, there just aren't any answers? There aren't any answers and it's a business model thing, but when it comes to the built environment, specifically, and things like smart contracts and stuff like that, I guess in that very American mindset is like, where's the money? Where's the business model and why do I care about all this stuff? How's it helping me do the job better? A reason why I'm living and working in Finland is because how do I take that concept of virtual

Fig. 4-4: A Gatherverse.
Courtesy: Amit Chopra of Iconic Engine.

worlds needing to make financial sense, because it has to be sustainable, and ideally, provide a comfortable lifestyle for those that are involved. But more importantly, how do we use the metaverse to improve the quality of life for the people that are doing whatever they do? And I don't hear enough people talking about that. As an example, there's this emerging thing called a Gatherverse that's going on which kind of highlights that, but again, I just haven't seen enough people talking about this (Figure 4-4).

PAUL: Nice to have you bring up such important points because it's about depth. So, I'm watching the NBA during the 2023 All Star Weekend, and they are doing some fun stuff with motion capture and AR technology. You can say, oh, that's the NBA Metaverse but the important points about what you're saying are what are the business models, financial models that make sense. But, the experience is actually the important thing. So, after the monetization of the NBA Metaverse, they are attempting to provide the stickiness factor of wanting to come back, because it's an experience that is about community. And what I'm finding is there's not enough discussion about balancing the wants and needs. This type of discussion can lead us down to a critical thing that I'm not hearing enough about. The FAANG (Facebook, Apple, Amazon, Netflix, Google) companies of Web 2, where they built their business models on a scarcity model, the scarcity mindset instead of abundance. And when you go into the abundance mindset, that's where Web 3 and these things that you're talking about, a nurturing environment, rather than falling back into that capitalistic, money is number one mindset. What are you valuing? At that point, the conversation of omission can be an important topic that this Gatherverse experience can have, that you mentioned. These conversations are currently not happening enough, but they must. Tony Parisi talks about the metaverse, where there are not absolutes, but inspirational growth paths. Where things are inclusive only to find that, if the metaverse is so inclusive, how is it that many people do not have access?

DAMON: Yes, inclusivity is saying that the metaverse should be accessed by mobile phones first. If you're building anything that's supposed to be inclusive in the open metaverse, and it does not start on a cell phone, you're being elitist. And you're targeting a very small, privileged percentage. So, that's where it'll be decades before someone in a developing country that currently has problems putting food on the table, water that is four kilometers away, is going to throw on some AR glasses to be in the metaverse. But they do have a cell phone.

PAUL: That's a really interesting topic about providing not just the device, but the connectivity so that you can enter these worlds. I thought that one of the big lessons from the book and the film, *Ready Player One*, was about escapism, because their reality was so stark and so bad, that their respite was to go into this collective illusion that we all say is a reality. I'd love to get your Crystal Balling about where are the danger points are as the metaverse starts to grow? And what we should watch out for as the built environment uses the metaverse? Because as an industry, we create the digital DNA of what will be a 3D representation in the digital realm. There's a school of thought that making the metaverse is taking your digital twin and "save as" onto a gaming engine and you have your metaverse. There are a lot of really great pieces of architecture and experiences that you can experience physically that fail when you put it into a gaming world. I always enjoyed the idea of that scene in Ready Player One where the two heroes go into the nightclub. And as they go in, the space acts and looks like a nightclub where there's gravity.

They can walk over to the bar and have a virtual drink the whole bit, but once they go on the dance floor, physics goes away. They can do amazing things floating in the space. What can we learn from the metaverse that could help with designing a better built environment in the physical world?

DAMON: If the metaverse turns into the movie *Ready Player One*, then I have failed and everyone else who is super passionate about this technology has failed. That to me, the use of the metaverse in *Ready Player One* is the equivalent of being hooked on meth. If you ever get the chance, there's a short film called *Uncanny Valley* that showcases VR. The film shows these people that are jacked into VR, and they're in VR in the same way that you would think heroin addicts behave because they're in this house, and they haven't eaten or left the house in days, and they're all in VR. So to me, *Ready Player One* is the nightmare scenario of when the technology goes wrong.

As for the built environment and the metaverse, there was an architect John Bruce who brought up a really good point. He is an architect designing a lot of virtual spaces and there was a conversation that asked questions like, why do we need stairs? Why do we need a roof? Why do we need walls? It is not like its gonna rain on your digital avatar. And then even if it does rain in your virtual world, you can have it stopped. But what was interesting is that they found that when they strip away those elements of stairs, walls, and roofs, most people's brains could not process this. They needed those elements to be able to understand the digital space. It was really fascinating to see the limits of what we can't let go of reality, even when we're in the digital world. For those of us who have spent most of our lives in the real world, we just have to have too many anchors, reminders, visual markers (Figure 4-5).

As far as designing the real world, the obvious use of a virtual world is having people go into a space before it's built, build that personal connection with it, seeing where things are. So, I am not sure how the metaverse directly benefits design but the concept of Mixed Reality is something I care about more because we already live in a mixed reality. I remember a guy, Henrik Vinson, that showcased a robotic arm that does drywall installations. He did one of the first metaverse conferences at Stanford in 2007. And then a couple of years later, we're sitting on a panel at another conference, and I loved what he said. And it stuck with me, because he was able to clearly articulate that we already

Fig. 4-5: Metaverse environment of architecture as music to create a series of results. Courtesy: Paul Doherty.

live in an augmented reality, The use of smartphones that provide a window into another digital world is one example. How does mixed reality add value to the built environment? How can mixed reality add value to design and how we do things? I don't have the answers, but just some of the things that I'm exploring and playing around with, like my having a focus with property owners, because they're the ones that ultimately have the money, the need and have more incentive to explore uses for the digital asset. In the case of a commercial building, how do my digital assets add value to myself and my tenants? I think that the assets that are created they can be used in these virtual worlds or they can be used in this stuff can be used to help provide metaverse-like experiences for tenants. I think designing more interactive spaces is a key. One of the things that I really try to drive home is that immersion, it doesn't necessarily have to involve you and a device. There are immersive spaces already interacting with people, like you walk into a room, the motion sensor detects that you're there, the lights turn on, the heating or something will change, and you never took your phone out. You just occupied the space.

As we see more of the smart systems that are able to better react to us in these mixed reality environments, the foundation will be the physical built environment because I've never seen someone doing virtual reality out in the middle of a field. I think there's more people that are doing VR, in cities and buildings and houses. And then that physical space is going to give you limitations to your experience, there are certain permissions around that physical space.

One thing that is not brought up enough of is the permission of the building owner for VR/AR to use their physical space for these VR/AR developers' digital stage sets. I think a lot of it has to do with the fact that you've got people who are pioneering this stuff who are more focused on the ingredients than the application. So, security permissions by the building owner must be discussed because a building owner not having a say in this process is just not cool. I can throw a piece of geo located digital trash in the middle of someone's office or their building or their business or whatever. And what do they tell Niantic (developers of Pokémon Go)? I don't want the Pokémon Go character Pikachu here? Imagine players of Pokémon Go trying to capture the game's characters by using the AR App and running into a ballroom that is hosting a wedding. This is disrupting my business. How do I turn that off? Because if I own this physical space, you can't come in and place a cardboard cutout of something in here without my permission, why should you be able to do that with your digital content digitally. Another issue is that AR devices, including smartphones, can scan the inside of buildings and homes in 3D without permission. Most AR programs need to do this to define the space for a game or other app functions. Who knows what a person can do with this detailed information? These are realities that the mixed reality metaverse is going to have to address and may delay the growth of the metaverse.

Another big issue is what constitutes digital rights management. So, a famous architect puts all their designs up in the metaverse and makes it publicly accessible. How is that secure? What license is that under? Does the owner know that this architect is doing this? The metaverse is exciting, but a lot of discussion, understanding, and thinking through the ramifications needs to take place now.

PAUL: The cyberphysical relationship, as the United States government officially calls it, is fascinating. In my home, we have installed solutions like Nest to control my conditioned environment. And Ring, so when my doorbell rings, I can be halfway around the world and I can see who is at my front door and have a real-time conversation with them. But you have brought up such important elements of the foundations of the metaverse, reality capture, and digital rights management. I wouldn't want to be the firm that gets caught up in the first case law about digital rights management, because it's going to set some precedent. When you start talking about smart contracts and capturing that data, that everyone agrees within that project team, in that community, that the data in that project is the truth. The data also does not end its value upon delivery of the building project and the contract stops. In a smart contract world, the data never stops working, especially for purchasing of materials. What a great way of doing a quick analysis to see if the building part manufacturers were actually selling us a 10-year asphalt shingle roof, because when we start doing an average, after 15 years of being in place in that neighborhood, it's running about six years. We can now have a different discussion with building product manufacturers because we have empirical data that has real serious connotations to it.

You said that in the virtual world you need visual markers in order to place yourself because our brains are wired that way. What gets interesting is a physical element, like a wall, door, or window may have a QR code that brings me into the digital realm while also being physically present, a good cyberphysical example. This process acts like a portal from the physical world to the digital world. In Dublin, Ireland, I visited the site for the new Dublin City University campus called Grange Gorman. As part of their construction process, they QR tagged all of their different systems, doorframes, windows, things like that, that you could geotag yourself and actually get further information. It was like peeling back an onion with layers and layers of digital information, this idea that we could be creating that cyberphysical relationship by creating these doors, these portals to further information from a pragmatic aspect.

I'd like to get your opinion on this idea of experience. Are we looking at portals that can bring inspiration? Do you see this as the next generation of where design could go?

DAMON: I think that's the new area that hasn't been explored. We've all seen the BIM and building lifecycle graphics. Where in any of those graphics where data can be used to enhance the tenant experience? It doesn't exist. I have experimented with this in mind by turning a Bed & Breakfast apartment into my

> own digital playground. I used reality capture tools and then asked the question; how do I turn this physical space into an LBE (Location-Based Entertainment) space? I have this digital data, I have this 3d model. Now, how can I use that to create virtual reality experiences that are custom tailored to this? How do I use AR to do this, so if a visitor/tenant is staying here, there's all these fun things that they can do. Thinking of the visitor/tenant use of the space beyond the physical is an ongoing experiment that takes built environment digital data and extends the traditional use to entertainment.

Building owners usually don't know that they have this type of data. And with the way that social media is going, most people access to built environment places like your house. If I take one or two selfies in your living room or I just come over to your house to play a game online and take a few pictures and post them on a social media site like Facebook, from the pictures you posted, they can use enough of that to build a 3D model. Facebook is doing this specifically for Oculus because they want you to be able to come into my living room and sit with me on my couch. And it looks real for you. It is real for me. And I didn't have to scan anything. I just took pictures of my living room over the years and posted them. So built environment data can be reused in ways we do not fully understand yet. Some cool examples of the integration of the physical and digital worlds are on Disney+. They had this AR app that when you held up your phone or tablet and they use the television as the marker, special effects came out of your TV, like waterfalls, and other cool features. I think that as we see more people thinking about new ways from a designer's perspective, we will see further advancement of the technology.

But that's also what frustrates me. I see all these ingredients that exist, and they've been available and there are people who are experts with them. But there are certain ways that they just haven't put them together yet. And that boggles my mind. I think it goes back to that they're focused on the ingredients, or the people that are trying to make decisions for how the metaverse is going to unfold or evolve, or all this stuff are the XR crowd, or the Web 3 crowd, or the AI crowd. We've have the wrong people leading the discussion, in my opinion, especially when it comes to the built environment.

Take for instance the recent trend of design firms putting their architectural designs online for the public to explore. Well, is there any sensitive information up there? Did you design a bank? Did you just show somebody a layout of how to rob a place or blow up a place?

We are living in an interesting point in time with no exact outcomes due to actions like this.

It blows my mind that with all of this uncertainty there are people calling themselves metaverse experts. There is still so much to figure out that these self-appointed experts of the metaverse are causing more problems than providing solutions. I want to see the metaverse be truly mixed, as its primary use should be for communications. The metaverse promises to help us be more present and connect with the reality that we're in right then and there. Do the opposite of *Ready Player One* where everything is dystopian nightmare where the metaverse becomes an escape to a not real, better life.

PAUL: Dystopian. What a word. The entertainment industry sees the future, they default to dystopian in so many movies and TV shows. When I see the future, I don't see *Blade Runner*. Who wants that? Hollywood is lazy when it comes to defining our future because it's very easy to do dystopian. Yet, it's also very easy to do utopian. But the reality in between is really hard to try and be that crystal ball. I know a company in Shanghai called DigiTwin that proposed to Netflix a similar solution to your Disney+ example. They also used your phone or tablet that launched a browser-based app that would be activated during certain episodes of the hit show *Stranger Things*. They embedded a QR code as an easter egg during a certain scene and the app would find markers throughout the room you were watching the show. When the monsters in *Stranger Things* were seen in the show, the monsters would not just come out from the screen, it came out from your wall, around your door, or through your ceiling. You could use your phone to actually be a weapon and actually fight the monster. This form of LBE brought up many questions to the player/user.. Are you a passive? An attendee where you just want to watch to see what goes on? Or are you participatory? Are you somewhere in between? It creates a whole new layer of entertainment as a direct result of that wall being somewhere on planet Earth with that screen that used to be your portal. Now, what happens when you extend that? And what happens to design at that point? We are talking about a future where rooms that you enter, not so much for escapism, but to interface as a communication device, almost, like entering the old phone booths.

One last thing I wanted to talk to you about, and it's about both of our pasts. I raised up on social media about a project that I was working on in 1996/1997 through Screampoint, led by O'Brien Chalmers, who is a legend in his own right, he does incredible 3D/multimedia work. He was friends with the people over at Planet 9 that got that San Francisco Giants Stadium project, and I happened to be the right place at the right time, because I was consulting with Screampoint, and went to these meetings, about the possibilities and how they were able to use VRML to actually make it kinda sort of work. It was imperfect, but they got it up on a browser at a time when AOL and CompuServe were considered the Internet. I published the screenshots of the ballpark in a browser, in my book *Cyberplaces* in 1997, just as an example of these virtual worlds in the metaverse. How did you know Planet 9?

DAMON: I've known David of Planet 9 for twenty plus years. They also published the San Francisco City 3D model and a variety of different cities, and at a certain LOD (Level of Detail) and I was leveraging that. Then I've just followed what they did, the whole idea of Planet 9 and what they were about, and the 3D world part of the Web is one of the things that intrigued me, it's what I'm doing today. To see their evolution of one of the first mixed-reality gaming platforms in 2010 to today's AI chatbot development, that's how I know Planet 9 is from their birth.

PAUL: Back in the day, they introduced me to developers in Silicon Valley that were putting together this virtual community called Alpha Worlds. It was a client application that you needed to download and install but used primitive Internet technologies to share a virtual space with others and communicate. I got involved there and actually took out a little homestead, built a little home, and could visit with other Alpha Worlds people. This was years before Second Life emerged. I never got involved with Second Life. Alpha Worlds was way early, a pioneering technology where I could hop into a community dome in alpha worlds, and literally chat with others through comment boxes coming up out of my avatar's mouth, and you'd hear sound, but it was like piped in Muzak. Web browser technology was still immature at that time but working through the Alpha Worlds application on my computer, I was blown away. Even at 2400 baud, dialup modem speed.

There's a lot of greatness out there that I believe that the new generations need to not reinvent the wheel but get on the shoulders of those giants. Do you have any advice that's been on your mind concerning the metaverse that could be like a crystal ball?

DAMON: One could have a lot of advice, but the biggest take away is that the metaverse does not exist yet, so there is nothing to learn from it yet. But there is a lot to learn from those that have been building out components of the metaverse for decades. If some says that have been involved with the metaverse for less than six to seven years, don't follow them. The people to follow are those who we stand on their shoulders, the shoulders of giants. Unfortunately, game developers seem to be today's leaders/experts in the metaverse, but they are actually the bottleneck. Just to get them interested in the built environment, we have to find some way of gamifying our need, which taints the result in being more like a game rather than the promise of an open metaverse. I would say quit looking for game developers as metaverse solution providers, but rather look into the built environment and find players of games and video gaming backgrounds for inspiration and know how.

Find those people in the industry that are interested in getting involved with interactive 3D technologies. Identify those people in your industry that are there from the younger generation that get it and quit going to gaming. That's just rude. That's like going to an architect and say, I want you to draw pictures of dogs. They draw but it's not what they're interested in drawing so to go to a game developer and say, make me something that is not game oriented, its not what they do.

Be involved now with the creation of the metaverse and understand that no one knows what it is actually going to be like. Not even the guy who coined the term, Neal Stevenson. It is cool that he has said in a recent interview that even he does not know how this is all going to work out.

We're currently building the metaverse for the next generations that are the next wave of architects, that next wave of contractors, I mean, it's hard enough to get contractors today to use iPads. The metaverse is planting that tree that you're not going to sit under. Had I known that when I got into this 23 years ago, I may have picked something that grew a tree a little bit faster, to be quite honest. My advice is just go out, learn from the people that have gray in their hair, understand what the value is. Don't get seduced by the ingredients, get intrigued and excited about what it builds, what is the dish, but then also look beyond that and say, but what is the what is the true value for humans? Beyond this just not the financial benefit? But how is this going to improve the life of the human being? You can be thought of and then so that's where I would just hope that people kind of begin with the end in mind, not be so Pollyanna or too excited about this optimistic future, be a realist then kind of reverse from there and then put themselves in that kind of experience of, if I'm the creator, what are the considerations? I need to have fun for the user, what are the considerations I need to have if I'm the owner? Because that's where especially for us in the built environment that combination of physical and digital space is important. I think that the Web 3 sense of data governance, data sovereignty, how do you give control back to the user with their data and things like that, that's another thing I would like to see more of. I think that we all should continue to build the metaverse that is for everyone with a sense of openness, and all inclusive. That's where I would like to say the metaverse should follow more the model of the Web than today's efforts by Meta, Roblox, Epic Games, etc. It's just the complete opposite.

PAUL: This has been a blast, thank you Damon.

Hugh Seaton – Data

I met Hugh as a fellow expat living in Shanghai, China. We would have these amazing conversations at a local watering hole and we became close friends. Hugh is one of our industry's top thought leaders, advisors, and entrepreneurs in regards to our industry. His past work with industry leading marketing, advertising and VR companies provided an amazing opportunity to start a company in our industry with the learnt knowledge from his previous ventures. His current venture, The Link, provides a solution that connects data from construction specifications to workflows and outcomes that I have not seen before. Since this data is so important to virtual worlds in the metaverse, I thought it was appropriate for Hugh to provide his vision of where our industry has been, where it is today and where we are heading.

PAUL: You came into the built environment by diving in with both feet after spending some time being on the toolmaking side of things for things like VR and the digital worlds that a lot of people are talking about. With what you're seeing out there in regards to equipment, software, accessibility, etc., where are we at?

HUGH: That's a great question. I think that people look at any technology, and they kind of draw a straight line from where it is to where they think it might ever go. But what they forget, is that with every adoption and every development process of a really useful product isn't a straight line. It's a logarithmic line. It goes like this. In the beginning, it's easy to do demos, and they keep getting better in a linear way. AI today is a good example of this. But this is true across the board, to MIT to really commercialize something, you go from that straight line of development to an increasingly steep line in terms of how much effort time and money it takes to make it really work. Because now you're dealing with all the little edge cases, all the little things that are specific to, it's a little bit lighter out, or I'm using it in this way, or I'm using it in that way. And the reason I bring that up is in, you know, 2013, everyone looked at what Oculus did and the fact that Facebook bought them and thought, wow, you know, this is real, it's coming. And this is if you can just imagine this getting a little better. They didn't know how much better it needed to be for non-tech people to be interested in it. And the reality is, it has to be insanely better. Because life is working for people like I've optimized everything I do around the real world, and to make the immersive things like VR work for you. It has to be enough better. There are a few parts to that. One is how good is the experience. And that's the tech needs to be great. The second thing is have we figured out things that you can only do in there. We talk about things like the killer app, but there tends to be uses that existed before a technology that are absolutely transformed by that technology. Certainly, the internet is a great example because it was cool. At first you played around on AOL or whatever. Then search came in all of a sudden, you could find anything quickly. Google really transformed the internet because it made it accessible. It also made it monetizable, which is also critical. There needs to be a Google for the metaverse. For all the fact that we walk around and live in 3D, we don't really think in 3D. Very often, when we're walking around in a room, it's perceived more like little islands of 3D, but we think more in a 2D way. And we spend all this time on screens, which makes it worse.

I've recently been reading about thinking in 3D. It has made me start to meditate about 3D, like I would have imagined a room in my mind in actual 3D. And it's so much harder than you think. We identify as people that are digital natives and non-digital natives, there's going to be an immersive 3d native and non-immersive 3D native. Nobody's been born yet as an immersive 3D native. This is not like people playing games. I think what will really be different, and AR can do this, because it puts things in the room with you in a way that that is much more interactive and immersive than then we're used to dealing with VR really well. I think this transformation will have impacts on how people think. Because there're analogies that 3D gives you that 2D doesn't and things like embodied cognition which is this idea that all of our thoughts come back down to what our senses tell us. There's a part of your brain called the somatosensory cortex, when you talk about far your arms light up, or when you talk about rough your fingertips light up. We relate even abstract concepts to the way we walk through and experience the world. So, I think down the road, there's an opportunity for humanity to think about concepts differently, because we are in an immersive world that is so immersive, for lack of a better word. And I think the way we represent information and the way we represent concepts can and will also become more immersive and more, more 3D. As an example, if Apple does come up with typing in the air in front of you, like they're saying they might, the way you type will not be QWERTY-style keyboard anymore, because why would it be, if you're going to have to relearn it, you're going to have to relearn it. This is a really interesting point, because now it makes concepts that we thought of in a flat map way into a spatially mapped way, and all of a sudden, it can be like synesthesia.

I think that we're barely at the beginning of the metaverse, but I think we've identified that there's probably something there. The tech isn't close, making the metaverse a super duper monitor for a while. We're probably a while away from really knowing how to use it in a way that unlocks truly new things. And they're probably things that already exist, that will totally change how we use them because of the metaverse.

PAUL: Absolutely. There's a series of designers and technologists that are from the built environment that are now working directly with the big tech equipment and software tool makers like NVIDIA's Omniverse, Epic Games' Unreal Engine and Unity. As they are developing tools and solutions, I wonder if these people are thinking in terms of the cognitive spirit of what you're used to, because your brain now is trained to think that when you're inside of an environment that there should be four walls and a ceiling or a roof. But why do you need a roof when it doesn't rain in the metaverse?

HUGH: Exactly! Why didn't you take it away? Why do you need it?

PAUL: But if you take it away, people freak out? At TDG, we had an experiment with autonomous vehicles in Qingdao, China. We used a digital twin methodology because we were trying to figure out the manufacturing of 20 m buses, 2 m personal pods, and a traditional 12 m public transportation vehicle. People loved riding in the 20 m vehicle as it reminded them of a train where you usually do not see the train conductor. The 2 m personal pod was a hit because it was like a ride at Disneyland. But people freaked out riding in the 12 m vehicle because they could not see who was driving the vehicle. They were used to seeing a driver in a traditional-sized bus. So, we put a fake steering wheel and hired drivers who faked driving and the results were astonishing. The passengers loved the 12 m vehicle. You are onto something when you bring the cognitive elements of human behavior into the discussion.

We do need spatial markers in order to have a level of experience or a level of engagement that allows people or gives them permission to start to explore. There is also a lot of energy about the metaverse not becoming like a *Ready Player One* type of escapism, which is like a form of drug addiction. Are we at a point where we need to learn what people are discovering, because at the end of the day, I don't want game developers to be designing the metaverse? There is a need for our industry to start getting involved so that we can at least have a say as these things are being shaped that will become elements in the metaverse. We know companies taking their BIM and putting it onto an Unreal Engine, and then publishing it as a 3D website. How can any of this be used to help improve the built environment?

HUGH: 100%. This relates to conversations about project delivery. If possible, it would be nice to delay certain decisions as long as you can until you can get a metaverse tool like VR implemented. This allows the owner and the person who is saying,

"I want the building to do this" to have them making certain aesthetic and experiential decisions as early as possible. The earlier you can put the VR solution in place the better because people aren't great at designing from an abstract. One of the ways the metaverse tools like VR has helped people is they'll make a model of something, and have the owner go through it. So, I've heard of this in sports stadiums, because they had them go through the corporate boxes that they sell. VR improved the experience so that everyone knew exactly what they were getting. Another example was a hospital. They had people walk through to understand what the operating theater would be like. They could stand there and feel if this design will work or not. VR allows you to get in there as an experience it as if you were in there and make intuition-based decisions, decisions that are judgment calls that require all of your centers certainly require a spatial sense. I think there's a lot of value in that.

I think people underestimate how hard it is to really imagine how to use their building space, especially if there's anything new, like innovative art or lighting in a lobby of a building. I worked on a lobby redesign project in Manhattan. They were doing some interesting things with shading and lighting and all that, and they spent a fair amount of money programming it to make sure that the experience was what they wanted. Using a FARO reality capture tool for the existing lobby, we created an immersive VR experience that highlighted issues like dimensions that were not accurate in the new design. We were able to walk through the entire lobby and showcase the experience with the new design that allowed a conversation about the innovations being designed for the space. In the end, the project was highly successful because VR provided the existing truth about the lobby and what the lobby wanted to be.

The metaverse has an incredible opportunity in urban centers where there's just enormous amounts of inventory that has to get reworked, and no one wants to take jackhammers to it if they can avoid it. The other one is new build when it's so experiential, that they want to make sure it's right.

PAUL: When I was growing up in the industry, I immediately was thrown into the deep end of the pool upon graduation from architecture school because I was vocation based rather than just the theory of architecture. I was taught construction documentation by hand drafting and writing specifications on a typewriter. Our school was known for graduating and the next day beyond the board's being productive. And that was a really important thing in place like New York City where I lived and went to school. All the architectural firms wanted to see was Asses and Elbows, because when drafting on the big boards, you'd have to have your elbows to prop yourself up to draw and your ass was always up drawing. In today's world, if you mention Asses and Elbows in a firm in Manhattan, I am sure HR would be notified and most of the younger generation would ask, what does that even mean? The architectural profession is working differently with new equipment and tools. The younger generation works sitting down with three to five screens getting their work done. This is how work gets done today.

I was surprised during my first experience of designing and delivering a new building. I was with a small architecture firm in New Hyde Park on Long Island, New York. I was the sole intern. The project was to design a new Salvation Army building in Mineola, New York. The project was designed really well. I was the guy that did all the elevations, the sections, the plans, hand drawing, and the written specifications. I was never so surprised to go out to the site while it was under construction and realize, "Oh, that's what it looks like." To your point, Hugh. I was trained to think in two dimensions. Even though I drew sections and elevations, I had to see it physically to appreciate what the building looked like in reality.

When I wound up as part of the team as a consultant that created the market leading BIM software called Revit, I had a little bit more experience about doing 3D. Revit provided an interesting twist to 3D in that I could parametrically change the model and subsequently the 2D construction drawings. My entire brain had to be rewired on how to design like that. Because all of a sudden, I was thinking and designing in 3D and it was hard.

That was an epoch type of moment where everything's going to change from this point forward. Today, as we begin yet another transformation of moving through space and time through metaverse technologies, we are faced with data flowing through these 3D models and how to manage the process. It's like our models are becoming alive and I want a more organic growth rather than having to deal with Frankenstein BIM.

It's going to be interesting to see how the designers of the 2D world who come from hand drawings and/or AutoCAD, who more or less gave up when Revit came, how the emergence of the metaverse changes what and how they design. I imagine some just want to retire instead of having to retrain their brains. Are we seeing a migration as BIM jockeys are having to learn new skills or get out of the way to a new type of professional that understands this migration of data-based geometries, rather than attaching data to geometry?

In other words, like with AI, what I'm seeing through tools like Open Ai's DALL-E 2, and seeing through tools like Midjourney, here are tools that are getting me about 60% to 70% there in my design process in a generative way. But are the BIM jockeys going to be able to keep up? Are we looking at maybe from outside the industry to start, a reimagining of what the design and construction deliverables are going to be, and then more importantly, the after effect of a digital asset, because it's no longer constricted to the timeline of a contract because the data can live on?

HUGH: When you talk about data, you take me somewhere else, and that is using data to optimize processes and do our jobs in a less uncertain way. And by uncertain I mean variable. The entire industry is based on the idea that each step is going to give them the next step, a step less than they needed, and it's up to you to figure it out. You can't really optimize very well if that's how you do good things. On the other hand, no one's paying an architect to design down to the bolts. Take, as an example, connection engineering. In the past, girder connection engineering used to be tons and tons of people with drafting tables that would literally design every joint, rivet, and detail. Then the US steel industry collapsed and they couldn't afford that anymore. There's a solution that came out of an experience with Bethlehem Steel in 1972. The founder saw the whole industry go through this collapse and decided to create software to assist with reimagining the connection engineering business. In traditional connection engineering everything was three or five rivets. So, every connection was either three or five, that was there was no two, there was no four. By utilizing BIM tools this solution users can now design right to the individual bolt.

Where I'm going with this is you're going to be able to do a lot of the design down to the bolts as an industry we collectively move forward due to data management. Another example is on the contractor and subcontractor side and their input into the design process much earlier than previously performed. This means that design becomes more and more augmented. So, you're not automating everything, you're just automating the dumb stuff. This all leads to you are giving people less autonomy, you're giving them a tighter scope within which to be problem solvers. They're not defining and then solving the problem. They're just solving the problem. So, it goes from they gave me this thing and let's figure it out to I have to put that there and make it work. And I have to do scheduling and the 10 other things that go into making a building delivery work. But the actual how does this thing get installed gets increasingly narrow in terms of how hard a problem it is. All of a sudden, once that's true, you can start applying data to these things and saying, "How fast did we do this?" While that's happening, you're getting more and more sensors, mostly machine vision in the workspace. So you can start to also really map how people are working, how the process is working, so on and so forth. As design becomes more and more Metaverse infused, whether it's because everything is driven by data or driven by geometries, the scope for problem solving narrows, you can have less qualified people doing things, which is nice because we have fewer qualified people, or people with fewer years of qualification, you don't have 20-year project managers anymore, you have maybe less than 10-year project managers. This framework also allows you to start to do process improvement in a way that manufacturing did in the 1990s. One of the things you're not hearing enough in this industry is Business Process Reengineering (BPR) that happened in the 1990s with manufacturers. When SAP, Oracle, and IBM infused things with data that made it possible to look at your process in a much more granular level, manufacturing went through a major transformation on every measure. Our industry is on the cusp of this similar transformation. The fact that you've created this data, this representation of the asset, and then all the data that went into building delivery and so on, you can now perform analysis, reporting and tasks to improve, which is the essence of BPR.

What will come next is people will start to ask how do we manage this data so that it's useful to someone later, either as a record for claims, which of course people want to do, or for maintenance, operations, or repairs. I think this will happen soon. It's harder to do than people think it is. Partly because so much of the data is coming from different organizations. The taxonomy issue is a major hurdle to automation efforts in our industry. In every area of our industry, people are calling things different stuff and classifying things differently. MasterFormat does help but it doesn't go deep enough to really get into some of this stuff. So, I do think that I'm answering a different question than I think you asked. But I think that that alongside this data is a representation of everything, which is a lot of what the metaverse is about, there will be a real opportunity for companies to get so much better than their competitors, that they start to really win. I am really looking at how they do what they do. That's how Toyota beat everyone for 30 years. It wasn't because they went and got some amazing tech. In fact, the US usually had better tech, because Toyota used their tech in a way that was humble, and was constantly improving, which is what Lean is all about.

I think that that's hard to do in a construction environment where everyone has so much scope of control that every day they make choices that are probably good ones, but they're not repeated choices. They're idiosyncratic every time to the degree you can constrain the scope of choice in the scope of problem solving, so that you can compare one situation to another. Now all of a sudden, you're able to use data to optimize things, and lower the requirement for experience and quality to still get a good job done, which is critical given we're never going to not have a labor shortage. But it also means you can get better and you can become the Toyota of the construction industry, in which case you can out price everybody with better quality. Think about what Toyota did, they had better quality, lower price, faster turnarounds, and more ability to do to do various things, that every measure was better, because their process was so deeply driven by continuous improvement. When a contractor does that it's over for their competitors. That's how you get to be a successful contractor in the cyberphysical world is because they do what Toyota did.

PAUL: What a great analogy! In your crystal balling, knowing that so much history repeats itself (meaning humans just repeat themselves) and as we start to improve our processes, as we start to sharpen the sword, and as we start to see the maturity of a new communications medium called the metaverse, what are you seeing?. My son is 12 and he seamlessly moves between the physical and digital worlds. He doesn't see it as two separate things. Are there analogies from other industries that you've seen that we could at least benchmark ourselves about? What do the built environment professionals do to help benchmark?

HUGH: It's worth noting that a lot of times change, like step changes can sometimes come not because of the opportunity, but because of the threat. And so, I think that asking someone to change how they do things because they couldn't be better, usually doesn't work. It's more like if you don't change, you're going to have to retire or whatever. An analogy I really like is the original in my mind, the original digital twin, which is aircraft engines. So, 30 years ago, or some crazy amount of time ago, GE would allow you to see what was going on in an engine that was flying over the Atlantic. That's pretty cool and it was all about sensors. And it was worth it because it was a US$10 million engine. This example is relevant to our industry because it fundamentally changed the way they monetize their product. GE changed their business model due to thinking differently about the data they were already capturing. They changed from being a company that makes and sells a product to a company that stands behind its product by offering a sale for 20% less than I used to sell this product to you, in exchange for a 20-year maintenance/service contract.

You could easily imagine a contractor or a developer saying, we're going to build this for you. And we're going to deliver this building for at least 20% less than anyone else. But we're going to maintain it for 20 years because we put it all in place.by investing in sensors to make sure that we know when stuff's going wrong. We know when there's variability in the electricity, we know when the heat gets extreme, and it has an impact on the windows. And as a result, we are able to make enormously more money as service providers, not just as builders.

I feel like that's one of the ways where you could see the unit, the multiplayer, the metaverse really mattering, because now you've got a maintenance division that may or may not be anywhere near that building. But they can go, they can go and look at the building, they can walk through it or skip through it as you do. But they can look at it, not as a set of dials or as a dashboard, but actually as a functioning building, and start to see where things are an issue and how they relate to each other. And you click a button and all of a sudden a maintenance schedule has been done for you. You can imagine that the building itself being an interface is an interesting way to leverage the intuition and experience of people that have done it all day long. Especially imagine now you've got a 70-year-old maintenance guy.

He's been doing this forever, and he really knows what to look for and so on. He's sitting in his home office with a Quest 2 Pro and is able to go to five buildings a day and just walk through them. And he uses real cameras sometimes because it's not so hard to do anymore. I think the metaverse is also going to have a real impact on people's ability in an industry where your body often makes you retire before your mind wants to. The opportunity to go and have a second part of your career.

I think the metaverse has a real opportunity for business models that we don't know about yet. Business models that are not out of thin air any more than Google's business models advertising. It's 200 years old, but they applied it in a new way and made more money they they're supposed to. Somebody once said that they had been the most profitable company in human history, if you aggregate if you add it all up just the amount of money they made, it's just added control based on versus their cost.

The thing that doesn't get talked about enough when people think about *Ready Player One*, and how wonderful it will be to be escaping and all that is the business models that are enabled by what Metaverse can do. And the other thing is, it doesn't all have to be fully beautifully immersive. It can be partial, it can be like an AR or pass through AR where if you think about what Apple's about to do, they're about to create something or about to watch something, that the quality of it is going to be such a step change from everything else that the cost is going to suck, it's going to be I think three grand, but it's going to be so clear and so helpful, so useful. I think it'll be a harbinger for the year or two later, when they have passed through AR that is also VR, I think that's going to really matter.

I think what's going to happen for us is to suddenly be able to say, "Mr. Contractor, we're going to invest intensively in AR technology as we're going to give people that are near the end of their of their career, the ability to hang out here and be remote experts." How cool for a construction company to say that I'm going to have a journeyman or even an apprentice on site and I'm going to have a subject matter expert be there with them telling them what to do to complete the task. I need a real guy there but having the option of using the metaverse tools like VR to emulate this situation, then we as an industry have new options.

One last thought, please imagine a building as a media player.

PAUL: Tremendous insights. Thank you, Hugh.

Cody Nowak – Process

PAUL: Welcome, Cody, my friend of many years and someone that I respect and listen to intently regarding the world of digital space and digital assets meeting the physical world. What are you seeing out there today?

CODY: If there is one thing that I'm definitely sick of is the term metaverse. Everyone's hyping it up is led by Meta/Facebook who is claiming that they are the metaverse. I have been attending events like Augmented World Expo (AWE) which I've been going to for years up in San Jose, California. It's actually the convention center that you and I first met up. This past AWE was interesting because everyone was just ripping into the metaverse, like everyone. If anyone said Metaverse during the presentations, there's all these grunts and groans in the audience. And then you heard a session that event lightly used the word Metaverse saying that dream of the metaverse was in the future, they were ripping into that. That word was being used by people that had no idea what the metaverse actually is, it's not here yet. It's not going to be owned by one entity like Meta. The metaverse is essentially another universe that where we're able to enter it through virtual reality, augmented reality, and it's available to everyone. It's like the internet of the world, but where we interface with any piece of hardware that is capable of transporting us into this new reality. And you're going to be able to access all the different contents of that new universe through anything. It's not going to be just one application. It's not one game. It's not one platform. It's a global platform.

What is really bothering me is the lack of discussion of a Built Environment metaverse. An industrial metaverse where we're able to access all this information of the different buildings that are being built in different locations, whether you're there in person, and you're seeing it through AR as buildings are being constructed. I think there will be different ways to access the Industrial Metaverse and it'll be interesting to see where that goes over the next decade.

What industrial metaverse is today, I'm not sure yet. But what I've seen since the release of DK 1 (Oculus Development Kit 1.0) in 2013 is that people are able to host their BIM in VR which is a really great fit for that tool. I think VR and AR are great fits for our industry.

Because we're creating our 3D content with BIM and that's what you need. In order to get into VR and AR. It just goes hand in hand, as soon as Oculus DK 1 came out and I was able to get hands on it, I knew that this is finally where we'll be able to have mass adoption of VR in the industry, where anyone and everyone can afford to have a headset. Around that same time BIM was really starting to be a thing more than what it had been, with adoption of BIM with many different trades. So, VR converging with BIM at that time was very timely. It's been interesting to see the adoption and the use cases that have come out of that. People using VR for coordination has been one of the biggest influences that I've seen VR have over the years. I'll give an example. When I first started working with CAVE (computer automatic virtual environment), that have been around since the mid-1990s, they were super expensive. The VR headsets used in the CAVE were also expensive and very rare. I found the quality of the interaction with the VR headsets just wasn't great, even with the high cost. Within a CAVE we had different modes to experience, 3D Active, 3D Stereo, Stereo, Stereo Scopic, and Stereo Vision. A CAVE also had a mocap (motion capture) system set up around the perimeter at the top of the CAVE. The CAVE itself is a big, empty, dark room that you walk into, where 3D images and other media are projected onto the walls, creating the virtual environment. Typically, a good size for a CAVE is 20 ft by 20 ft, by about 14 ft tall ceiling. This size gets all the Hardware and Technology in there. The mocap system that I mentioned is tracking head movement from the individual. They're now being tracked in 6 degrees of freedom. This provides a 1 : 1 ratio of one step in reality is equal to one step in virtual reality. As you move around a CAVE being that one person in that space, you get the sense of immersion through the 3D projection that you're looking into. And that 3D stereo mode is like Quantum Medium, it's the same technology as they're powered by battery where they just flicker really fast. This fools our brain into seeing 3D where if you took off your VR goggles, you have this blurred vision of a flat 2D image. The first time I saw this tech being used for our industry was in the mid-2000s when an architect and an owner walked through a building design in the CAVE as part of a pre-construction process. The BIM was projected into the CAVE that allowed the owner to better understand what they're actually getting and what they're paying for. This was a really big deal at that time. Although the price point for the CAVE and the VR walk through experience was in the hundreds of thousands of dollars, they found great value and they claimed that it was worth the money that they invested, which really left a lasting impression on me.

Typically, from my experience, projects don't usually have a budget line item for an experience like the CAVE. About 10 years from that first CAVE experience, companies started to ask me to build a similar system to the original CAVE. This is how my company CUBE began. CUBE stands for collaborative, ultimate building environment. And the BIM CUBE is the product. What I did was take the CAVE concept and what I'd learned from working with CAVE systems and hack it into the essential tools that would be needed for 3D coordination. I also wanted BIM CUBE to allow the project team to be both virtual and inside the BIM CUBE simultaneously while working with the projected BIM project model in 1 : 1 scale and have the ability to zoom into specific areas of the model at any scale you would want. I partnered with VR headset manufacturer HTC and used the HTC Vive as our VR goggles.

One of my customers was a hospital and we used BIM CUBE to coordinate and verify the proper design for different areas of the hospital. Using the VR goggles inside the BIM CUBE, we had a nurse, a doctor, and the architect virtually enter the exam room. They were all free to walk through the room. The doctor saw a design flaw in VR with the sharp needle trash receptacle being above a trash can. If a needle accidently missed the sharp needle trash receptacle and wound up in the trash can, this would be a biohazard. Using the VR tools, the doctor was able to grab the needle receptacle off the virtual wall and place it on another way out of harm's way. The movement in the VR model of this receptacle was automatically updated in the original BIM and issued the 2D drawings as a revision to the design, all automated by the BIM CUBE. This streamlined way of making changes became a superior process to the traditional process of note taking, updating the drawings, etc. In a complex, high-performance building like a hospital, there were hundreds of minor changes that the BIM CUBE took care of, streamlining the process. All taking place before anything is built or installed.

I always tell people that everyone's experience in VR is unique to themselves. I've put thousands of people through VR and AR, there is some common feedback of what people are experiencing like I you used the headset too long, a feeling of being nauseous could occur while most others just have a sense of awe and are inspired. But each person perceives VR differently. I once put a VR headset on a guy and he just before I could even strap it on he fell out over on the ground.

Luckily, I grabbed and saved the headset. The typical experience is a wow factor people feel the first time they enter my virtual worlds. People enter my 40 ft by 40 ft BIM CUBE, and they see the projection of something cool like an animation playing on two of the walls in a corner of the CUBE. The ceiling is about 12 ft high. There are typically 10–15 people who can enter the CUBE at a time, which means the projection is the immersive experience, not everyone wants or needs to strap on a VR headset, including myself. When I put a VR headset on it's a "get in and get out" process. The immersion in the CUBE begins when the people begin to walk toward the corner of the projected image. About the time people are in the center of the CUBE, the wow factor occurs. This use of projected immersion for BIM rather than using VR goggles all the time creates a valuable experience as people can communicate face-to-face while having a semi-immersive experience simultaneously. This allows the project team inside the CUBE to be able to read people's body language, look at them in the eye, conduct discussions naturally and in a familiar environment that they are comfortable in. In the case of CUBE, technology enhances the decision-making experience rather than the technology being the center of attention.

A big reason for not going full VR headset in the CUBE is the headsets are not quite there yet, but they are building toward a better experience. Meta (formerly Facebook) has Oculus Quest (Quest 2 and Quest Pro). In my opinion they are overpriced but very functional as an all-in-one headset. They come with cameras that look in at your face, your facial expressions, and soon to come eye tracking. Microsoft's HoloLens and Magic Leap are also doing this. These units also do hand tracking, giving you the option of not needing hand controllers. Hand tracking will make hand gestures the new interaction with the metaverse. With eye tracking, these units will be able to manage your gaze control, meaning they can track what you are looking at and you will be able to select objects by objects by looking at menu bars and selecting. This tech will also with your avatars as you'll now get spatial facial recognition where it will show your emotion, going to the body language that I was talking about earlier, and then also the eyes. It's not quite there yet, but everyone it seems is working on this.

And it's a known issue for the metaverse for any kind of co-presence, where you have one or two or more people, you're getting that more of the body language, and it aligns with how that person is talking, what they're doing, the interactions they're having. So we'll get there. It's not there yet. But for now, in something like the CUBE, you're in the virtual environment, but you're there in human form, as well. Once you get a team in the CUBE it becomes an interesting process to see how they react. Sometimes a stereotypical construction worker that's a bit rough around the edges and has been around construction his whole life but knows his stuff. He's Top Dog and he wants everyone to know it. They are the type to enter the CUBE a bit more reserved, staying in the background. After the first five minutes the wow effect begins to wear off and it's time to get down to business. The BIM, documents, and files get uploaded through a thumb drive and the session begins with the Revit model on one wall, a 3D walkthrough on another wall, project documents on another wall and I find the project team members begin to walk up to the projections and begin to touch them. There is usually a VDC person who can make the changes to the BIM in real-time based on the conversations and agreements the team makes during these sessions. It's around this time the Top Dog comes out from the background and begins to lead the discussions.

The CUBE process is super quick in comparison to taking notes, making red lines on drawings, marking them up, and sending them out to the BIM department. Hopefully they're understanding what the change is, or if that change is what that person's that's making the markup actually really wants. The CUBE is a game changer because it saves tons of time and money.

The CUBE is also used as a way of "flashing curb appeal" while on the job site. When clients stop in, or trades, or building product manufacturers, the CUBE allows the project to shine by making the project sexy. People really enjoy that.

Paul: The CUBE sounds like a great way of using metaverse components for use in the real world, today. Some people have called exactly what you just described, the industrial Metaverse, what do you feel about that? Because are we getting closer to a definition by creating those vertical metaverse solutions for healthcare, law, education. etc.?

Cody: That's a great point. I see a lot of different startups in the industry that have been focusing on specifically this for our industry. The vast majority of them are just taking a 3D model (BIM) model, allowing you to open it up into the virtual environment that provides a real-time 3D walkthrough, where you can then put on a headset and view it in VR. Some of these services may use a mixed-reality approach, like

HoloLens, and take the model out into the field and have an overlay for QA QC. That's really what the solutions are right now. What is emerging are new headsets that are specifically focused on what they're calling Engineer Grade AR that provides submillimeter accuracy with a headset. If you're operating at submillimeters, you're within all tolerances of all the different trades. Because if you're in framing your tolerance can be an 8th to a 16th of an inch for laying track, which requires accuracy, but at submillimeter, that's a whole other level. So, until we get the next generation of headsets and wearables, a lot of these startups are doing the same thing over and over, which is taking your 3D BIM, importing into a gaming environment, take it to a job site, overlay it out on the job site, and then do some sort of annotation that gets passed back through to other users, like in the BIM department. Maybe they're looking at what you marked up in 3D on an iPad which is good, but there are so many other things out there that this technology can do. Like mixed reality to improve the process steps for construction. Because that process especially when something is repetitive, like shop drawings or 2D construction documentation, and now be able to have something like a QR code on individual piece of equipment that's about to be installed, magic begins to happen. I did this previously with framing, you scan a QR code, and then it shows you where to go and you snap in your bolt in your framing studs. Quick and accurate. So being able to just have it where mixed reality is showing, what's the next thing, where's the next thing go. That's where the handoff is to an individual trades person that knows where that the studs need to be 16 in. apart or 24 in. apart. Or an electrician that is putting in a duplex outlet that is X amount of inches off the floor. If there are conditions that don't allow these types of typical installations, that's when mixed reality can assist them, as mixed reality gets them close enough and within tolerances. This really improves the speed of the installation because they're not going back and forth. And measuring twice cut once, as the old saying goes, because it they know exactly where it's at. It's been shown in 3D in one-to-one human scale and is what we understand. It's what we live, breathe for, as we speak right now, we're in one-to-one scale and we understand what we're looking at

> what our environment is, but when you take that back to a 2D environment with symbolic representation of what the 3D environment is, people can get lost. In my career across 20 years now, I've seen so many people in the industry that can't read 2D construction documentation, like architectural drawings. They have no idea what they're looking at even a floor plan. I guess I lucked out because I started in architecture and I drafted for the first 11 years of my career, like that's hard and heavy, I was doing a few other things but like, that's what I now live and breathe you know, and, and that goes for architectural detailing, 2D detailing, as well, which can really get people lost. Being able to finally have the 3D represented is an aha moment where anyone can start to understand what they're looking at. Because ideas come to life in 3D. And whether it's scaled or one-to-one human scale, it becomes a language that is already known because going back to everyone's born into 3D, into this one to one human scale of the world. It's really incredible to see the lightbulb going off when I bring people into a virtual environment.

This is why there is great anticipation for the industrial Metaverse. People are able to understand it instantly. When we implement a BIM CUBE onsite on a project, we'll have architects come in and look to get into VR because they don't have VR in their office. They may have BIM and VDC guys that are doing the modeling and creating the construction documentation. Inside the CUBE, they look at some of their details in 3D and say to themselves, hold on, this is completely wrong and begin the process to get it right. This is a good example of the beginnings of an industrial metaverse for our industry. What will our industrialized metaverse look like? Who really knows until we get there.

What is happening right now is that there's all these startups announcing that they are metaverse this and metaverse that. And of course, not all these startups are going to survive.

Technology needs to be developed at the industry level of AEC professionals for the industrialized metaverse to be able to find the challenges and then create the solutions around those challenges. Because just creating some technology doesn't mean that it becomes a tool that is used in our industry. And that's the age-old problem with software development or technology development being developed in a vacuum. You need our industry's input in order to be successful.

Which also means our industry must be willing to communicate with these really smart people who want to know what the target is so they can nail it.

As the industrial Metaverse emerges in our industry, we know its definitely coming, it's being built. I think that as the products mature and get into the hands of the professionals, industrial metaverse tools will emerge that provide great value. I think that that's where it's really going to take off. But it's going to be a little while before we really see it take off.

PAUL: Cody, I'm all about this as a snapshot of a moment in time of a maturing new medium. I would definitely love to have you stop in as we continue on the journey, because we're going to be creating our own virtual world and Metaverse based on the book. So, we'll have you as an avatar, hopefully a humanoid with me, that will continue our conversation. Thank you, Cody.

Arol Wolford – The Industry's Future

I was introduced to Arol in 1994 when he was the owner of CMD Group, an Atlanta, Georgia-based company focused on construction data. I had just presented at an industry conference in Atlanta that talked about our adoption of the web browser as our project management tool in our design build firm in Westbury, Long Island, New York. Web-based project management became a big deal and I began to consult with Arol and CMD. Over the years, Arol would take an interest in the latest thing I was working on and we became close friends. Arol published my first book called, *Cyberplaces: The Internet Guide for Architects, Engineers, and Contractors,* in 1997 while also working with me on numerous start-ups for our industry, with the most famous of them being Revit that was sold to Autodesk in 2001. Arol is currently my partner in TDG and I thought it would be good to end this book with a look back to where we have come from and where we are heading from the viewpoint of an industry giant.

PAUL: Welcome, Arol! Being such a forward force in the industry your entire career, especially on the technological side of things. And between Revit, and all these other tools that you have brought to reality, one of the things I'd love to get is your opinion on creating the building blocks toward the metaverse. In other words, where is this really going for us?

AROL: You bet. Well, it's always an honor to be working with you. And I know we've been pursuing a lot of the components to create the metaverse for a long time. My guidance, whether it is digital twins or metaverse metaphors, I see them very much as an overlap and extension of one another. But my guiding principle is what I call FACES. Whether it's a digital twin or the metaverse, I think FACES applies to both. It's something that gets overlooked philosophically and physically. I'm a biology major, so I'm guided by that bias. I think this concept of FACES comes natural to me, as a biology person who is focused on genetics and data. I like to set my data up in a biological type of way. And when I look at our industry, I use the acronym FACES. What does that stand for?

First is F, the Function of the building. It is distressing that we don't look at the function well enough. My example would be that Ireland did a great thing in the last seven years; their hospitals used to have no true OSHA's rate of people getting sick when you go to the hospital. Like the United States, approximately 15% of the people who go to the hospital were getting sick at the hospital. A recent study was performed to address this issue. One result of this study was that our industry must reconsider the basic function of buildings, like a hospital, and address things like quality-of-life issues. As I talk about FACES, what I'm really talking about is holistic architecture, just like holistic medicine, you don't just focus in on one variable. The study also concluded that Irish hospitals needed to raise their level of cleanliness, which can directly correlate to the function of the building. By addressing these issues, Irish hospitals now have less than 5% of their people getting sick inside of their hospitals.

The next acronym letter is A for Aesthetics. I think architects are genetically encoded with a really neat aesthetic gene. You normally don't have to talk about the need for aesthetics with an architect. That's one of the reasons I'm an architectural groupie, because I don't have that natural gift of aesthetics. My wife does. It's probably one of the reasons I was driven to this whole concept of taking a building information model (BIM) and transforming the BIM into a virtual information model (VIM) to allow the aesthetic expression to be better communicated. The concept of Autodesk's BIM solution called Revit is very sophisticated. I talked with Professor Chuck Eastman at Georgia Tech and you, Paul Doherty, about putting the Revit model into a gaming engine so we can better use the BIM with more people having access to the digital model to improve communication in the industry. That is our concept about VIM. Now, how do I tie that back to aesthetics? Down here in Atlanta, I went out with my friend, Rodney Cook, and we created and developed a monument on 17th Street. It was part of a US$2 billion real estate development project where they took an old steel mill site and developed it into a nice neighborhood of Atlanta. When Rodney went to show developer the design of the monument on 17th Street he showed them these beautiful drawings, and the developer, Jim Jacoby, kind of glazed over. And I know the feeling. Rodney had these beautiful drawings; he was explaining things. It was all in his mind. Poor Jim Jacoby is like me. When you're talking about aesthetics it doesn't naturally come to us. We need to see the design aesthetic differently. That was the opportunity to create VIM because Jim Jacoby just said no to us about giving us this piece of land, even though we're going to donate this beautiful monument there. We had the Revit model; we had the gaming engine and we had the need to better communicate our design aesthetic of the monument to Jim. I remember contacting and working with you on this project, Paul. By combining these technology components, we created a virtual world of Atlanta that resulted in a 90-second video to showcase our design aesthetic. We showed the video of flying down 17th Street and arriving at the monument to Jim. He was able to see the vision contextually and aesthetically. Jim said, "Arol and Rodney, give me that video and I'll give you the land. Now I know what you're talking about!"

Next letter in the FACES acronym is C for Cost. I own this business called R.S. Means that publishes the data of cost for the construction industry. I'm into data. I know it's boring for architects. I know they hate costing, but you got to figure out the costing, you have to have some awareness. When you tell the owner, optimistically, I'm going to design that hospital for US$80 million. And everything comes in and everybody knows it's going to be more than US$100 million, that is not a good thing. And what do they do to take care of that? Value Engineering, which is neither value nor engineering, it's just cut the costs, because the architects oftentimes don't take the time to come up with the appropriate cost. So, if you're going to follow FACES, you better figure out the cost, it will come up and bite you. And where bites us right now is we keep value engineering, which means we take the original energy efficient US$800,000 chiller unit and swap that out for a US$500,000 chiller unit that will pass code. A few years down the road, that energy-inefficient unit is polluting the planet and wasting a lot of energy and money. We feel that by using VIM and virtual worlds, designers can see the cost of their design decisions while they are designing, providing them guiderails to deliver a building that is within reasonable range of the project budget.

Next is the E, for the Environment. I think we all agree that the environment is critical. I remember in 2005, when the gentleman from Exxon shared with us, Paul, that 30% of the world's energy is being used by buildings, and 42% of CO_2, it knocked us on our backs. But it got us going, the environment is so important and with tools like VIM that can highlight environmental issues before the building is built, we are on the right track.

The last letter of FACES is S for Safety. Most large general contractors actually do a really good job with safety, but you can't be too safe. The use of digital twins, VIM, virtual worlds in the metaverse provides useful solutions for safety, like simulations and recommendations for operating a safe environment for your project.

If you're going to create something that's worthwhile in the metaverse, you better be thinking about FACES, function, aesthetic costs, environment, and safety. I think those are really important guiding principles.

PAUL: I think that that is just wonderful for the readers to understand the practicality of FACES both in the physical and virtual worlds. It does come down to communication and the 3D visual communication sometimes is easier for some people to understand. I know some architects that literally think and design in 2D. There was a huge shift in mindset when we went from hand drawing to CAD. But then when we went from CAD to building information modeling in 3D, we seemed to have lost a generation. Some don't know how to design in 3D, they can think about it sometimes, but they may not know how to design it. This brings a whole new conversation about design, both IRL and digitally. There is an entire generation of BIM coordinators that are highly talented and really good with the technology. Some of them are design architects. But the design profession has generation generations under the same roofs of their firms that are all working with different tools with different ideas of design. There is the hand-drafting generation, the CAD generation, the BIM generation, and now we have the Gamer generation. Communicating the design intent of a project means that the conversations from the very top need to happen with conversations with the people that are just entering the profession, as each has something to learn. It's no longer that an intern is listening to a senior architect because a two-way conversation needs to occur. Using metaverse tools like VIM provides a bridge of understanding that allows the different generations to talk with each other. As you are working with most of the iconic architecture firms in the world with VIM, where do you see in your Crystal Ball all of this going? Is it a metaverse tool like VIM that will help the generations come together and exchange information both upwards and downwards?

AROL: Well, it is, it is exciting. And I think it's kind of fun to have a biology major, like myself, get together with the architect yourself. And it's reminiscent of the past 25 years that we have spent together in the Design Futures Council where we get scientists from Dr. Jonas Salk's perspective coming in, and psychiatrists, because his son's psychiatrist, interfacing with the architects. I think that is the really beautiful quest for the future, that we're thinking more holistically about things. And if we could be humble enough to think holistically, we're going to serve better, and we're going to create something that that's greater.

So, we can't look at ourselves, as let's look at holistic medicine, what it's done for the doctors, the doctors used to be these master surgeons, very bright, very skilled people, they fix up your heart. But you know what? You're back a year and a half later because you're eating crappy food they didn't talk about maybe you need to cut back on the meat. Maybe you had better start to exercise as well. No, they just fix the heart. We, in our profession, cannot think as an architect or think as a constructor, or think as an owner of the only things that I'm doing. We really have to push toward more holistic thinking. Now the beauty of the metaverse is that it is holistic in nature. You do have data, obviously, you do have the visual, but the metaverse is looking at things in a more holistic fashion. And you know, you mentioned physics. Yes, I think it involves physics and biology. It has great design. So, this concept of holistic thinking is critical for the metaverse, especially if we can overcome some of the real dangers that you're pointing out, Paul. I really, I really liked the concept of bringing people together.

And you know what, if construction and design is seen more holistically, we're going to get more people wanting to enter into our industry. I have been on the board for Georgia Tech's School of Architecture over the last 20 years. Twenty years ago, it was 80% men with very few minorities. It's completely different now, I think 55–60% of the architectural students right now are women, which is a whole other beautiful breakthrough. There are real breakthroughs because this holistic environment has people who are thinking differently. They are approaching things differently. Frankly, they are thinking more holistically, They're just new minds. I think we have that same opportunity to not look at construction just as architects just as constructors, building product manufacturers, all vital. But we need a lot more breadth that the metaverse is going to be bringing us. We do have an opportunity to bring the forces of function, aesthetics, costs, environment, and safety all together. That's a big challenge. But I think it's an exciting challenge. I like talking to my grandsons and granddaughters who start out playing a creative game like Minecraft and continue on their creative journey by playing Fortnite. They have tools to create things. That's exciting to me.

As you can tell, I get excited about holistic architecture and design. And I think the metaverse is a wonderful new technology that can truly improve the built environment. I think the Smart Cities concept that you introduced Paul is very holistic in nature and will also be assisted by the metaverse. I know it was inspiring for me, and you're early on with that. Gosh, my wife ended up basically getting a PhD in holistic architecture. That was literally her thesis that had quite an effect upon me. And architects like John Carl Warnecke and Fay Jones that we both got to meet truly thought in a holistic fashion as well. That's fantastic.

PAUL: Always a pleasure to reminisce while thinking about the future with you, Arol. Thank you.

Conclusion

Defining the metaverse in any strict, regulated method does mitigate its effect. The romantic, unrestrained expansion of this wild grab for virtual real estate, often described as a "Gold Rush" (Ye, 2021), relies heavily on imagination, not quantifiable data.[2]

[2] The Metaverse as Virtual Heterotopia, David van der Merwe, 2021, www.socialsciencesconf.org.

The existing state of the metaverse as one that is emerging further clouds its definition and implementation for our industry and the world at large. Nevertheless, the metaverse has crossed the mind-share chasm of acceptance and there are many pioneers willing to put their brand into the leadership position. The leaders of this metaverse homesteading process will be in position to grab a lion's share of monetary transactional gain but will probably miss out on what the soulfulness of what the metaverse is positioned to reflect for the world.

At the moment, the metaverse is a collective illusion of a 3D world where people can communicate, collaborate, and cooperate that transcends reality. This collective 3D illusion also has the power to enable digital transformation in the formation of an industrial metaverse, focused on the improvement and enhancement of the built environment and other industries like healthcare, education, transportation, and a myriad of others.

It is unlikely that "the metaverse" will abruptly appear as a fully formed product or market, it will instead deploy incrementally over time as technical innovations and business models are proven, and interoperability standards are created to enable scaling to pervasive scale. Many companies and standards organizations are investing today in initiatives to make the building blocks that will be needed to deploy the broad vision of a spatial web, while reaping the benefits of intermediate use cases that become enabled. As with many disruptive changes, the metaverse will probably take longer than we expect to materialize but will have a larger impact than we expect when it does.[3]

The metaverse could emerge as a self-sustaining and persistent virtual world that co-exists and interoperates with the physical world with a high level of interdependence. As such, the avatars, representing human players in the physical world, can experience cyber–physical activities in real-time characterized by unlimited numbers of concurrent users/players theoretically in multiple virtual worlds.

In a perfect world, the metaverse can enable interoperability between platforms representing different virtual worlds (i.e. enabling users to create content and widely distribute their content across virtual worlds). As an example, a player or developer can create content for a game (Minecraft) and transfer this content into another platform or game (Roblox), with a continued identity and experience.

[3] Neil Trevett, President, Khronos Group and VP of Developer Ecosystems, NVIDIA, www.versemaker.org.

To a further extent, these platforms must connect and interact with our physical world through various channels, user's information access through head-mounted wearable displays or mobile headsets, contents, avatars, computer agents in the metaverse interacting with smart devices and robots. I am sure I have not listed all the interconnected devices and software solutions that will bring the world of the metaverse to our current reality, but I am sure that this transformation will happen sooner than we anticipate.

I look forward to interfacing and communicating with you online as the metaverse emerges from its current embryonic state. I wish you much success in your personal and professional lives with the emergence of the next great technological innovation, the metaverse.

Index

E

F

G

H

I

L

M

N

O

P

R

S

T

U

V

W